MARINE CORPS DRAFTEE

Eddie Nickels Parris Island- 1966

MARINE CORPS DRAFTEE

A VIETNAM ERA DRAFTEE'S PERSONAL EXPERIENCES OF

PARRIS ISLAND AND INFANTRY TRAINING REGIMENT

EDDIE NICKELS

MARINE CORPS DRAFTEE

EDDIE NICKELS

ISBN 978-0-9886933-1-9

Printed in United States of America

CONTENTS

FOR MY WIFE, CHILDREN, GRANDCHILDREN, AND GREAT
GRANDCHILDREN, WHO PROVIDED ME WITH THE
INSPIRATION TO WRITE THIS BOOK

Wanda Nickels, Steven Nickels, Alisha Bowling, Jeffrey Nickels,
Tyler Nickels, Ryan Nickels, Kimberly Nickels, Jordan Lucas,
Summer Lucas, Carly Bowling, Dylan Nickels, Daniel Nickels,
McKenna Nickels, Tayvin Nickels, Brooklyn Nickels, and Carson
Lucas. Also for Ricky Bowling, Michelle Nickels, McKayla Lucas,
Kayla Nickels, and Felicia Hamilton

Acknowledgements

Thanks to my wife for her help in preparing this book and for being my "sounding board" when I had an idea or phrase I wanted to incorporate into an incident or event. Her advice and suggestions were invaluable to me.* A special thanks to Michelle Nickels, Jeffrey Nickels, and Steven Nickels for their technical assistance in the preparation of this book.

The members of Platoon 381, S Company, Third Battalion, Parris Island, South Carolina will always be part of my memories of Recruit Training. My story is their story too.

The members of Platoon 381 include the following;

Ronald Allen	Maurice Burton
Thomas Allen	George Butler
R.A. Andreoli	Burton W. Butts
James Bailey	J.R. Carrington Jr.
R.R. Berardino	R.C. Carver
Terry A. Bishop	James F. Charis
S.M. Breneisen	M.J. Christmas
Hans R. Bronner	Aaron Cooper Jr.
Conrad Brown	Robert S. Corwin

ACKNOWLEDGEMENTS

Donald Brushi	Richard R. Cote
Gary R. Crawn	Ricky J. Labbe
Joseph W. Davis	Robert Lafleur
Ralph G. Desena	Thomas Laneave
Gene B. Dixon Jr.	Albert Lausier
Dennis Domagala	Edward G. Leite
Paul E. Doty	Richard A. Levy
Donald W. Drake	Richard A. Lewis
Ronald Duva	Willie Lovette
John R. Farone	James E. Lux
Darrell Fleming	Edward Maddox
M.P. Fredrick	P.M. Marciano
R.L. Gamberale	James McArthur
Glenn A. Graf	Michael McCaw
John Habermann	Robert L. McRae
Donn Heizmann	Francis Miller
Carroll G. Hill	J.N. Miller
Edward Hoffman	Robert A. Mosher
Norvell Jones	Henry Mumme

ACKNOWLEDGEMENTS

John E. Keating	Eddie G. Nickels
James M. Kinder	Joseph P. Owens
Jeffrey A. Peck	Myron L. Yelverton
Freddie A. Roach	R.M. Englehardt
James Robertson	D.H. Dale
Edward J. Ross	D.J. Giancarlo
R.S. Schrader	D. Arnone
S.W. Shifflett	C. K. Carey
F.J. Sledjeski	N. Kantargi
Thomas Staples	P.M. Carmody
Raymond Stevens	R.W. Hawkins
P.G. Strosnider	W.V. Hartness
C.W. Sunley	R.D. Fraze
Alex J. Terlesky	R.H. Jones
John Therlault	D.L. Lewis
William Todd	Pvt. Monte
Jamed D. Tussing	Pvt. Watson
T.D. Walton	T.R. Wekenman
Jerald Weaver	John L. Welch

ACKNOWLEDGEMENTS

Nathan R. Weiss

Third Battalion Platoon 381

Commenced training   Completed training

March 26,1966   May 20, 1966

Lt. Col. G.L. Lillich   Capt. H. L. Haley

Battalion Commander   Company Commander

1st Lt. E.E. Record –M/Sgt H. Massey –   Gy/Sgt. Kane

Series officer   - Chief Drill Instructor - Series Gy/Sgt.

Drill Instructors;

Senior Drill Instructor-S/Sgt. T.T. Lister

Junior Drill Instructors- Sgt. J. C. Todd, Sgt. M Gaitan,

Cpl.M.W.Henry

INTRODUCTION

A PLACE CALLED PARRIS ISLAND

Although my story encompasses the complete training regimen that goes into the training of a basic Marine, I have put more emphasis on boot camp and Parris Island and the eight weeks of training I received there. No one that has ever stood in those yellow footprints and trembled as Drill Instructors screamed in their ears and shouted out orders impossible to obey will ever forget their experience. The change we underwent there is forever and will never be forgotten.

Those Drill Instructors have the responsibility of turning sometimes lazy and slothful citizens into Basic Marines and they have very little time in which to do it. (Especially during wartime when the training schedule is often shortened to meet manpower needs.) This leads to extreme stress not only on the recruits, but on the Drill Instructors as well. The results could be hazardous to one's health, especially to the recruit's well-being. I have tried to soften the events I and others experienced in boot camp but I have also tried to be perfectly honest about the trauma I experienced along the way.

You can talk to a dozen former Marines, (there are no Ex-Marines,) and you will get a dozen different experiences and many different opinions of the training on the Island. This story is mine and mine alone but the incidents are true and they are what I experienced as I strived to become a Basic Marine. If there are any mistakes in the book they are mine alone and are due to the passage of years since I went through Parris Island.

In writing about my Drill Instructors I never use the abbreviated letters of "D.I." because we were never, ever, allowed to use those words during boot camp. To do so would bring down the ire of a Drill Instructor upon the head of the offending recruit. That's why it is so hard for me to use the abbreviation even today, forty seven years after the event. Throughout the book I have spelled those words out even though I was tempted at times to give in to the abbreviation.

I have used military time instead of the civilian clock and have used many naval terms and words when appropriate. Calling a door a "hatch" is something I still find myself doing occasionally and every Thursday I think of that day being "Field Day" when a thorough cleaning was performed in our barracks and living quarters. I still often call my bed my "rack." When using civilian words instead of nautical terms would result in harsh punishment, you quickly learned to use the nautical word instead. It's not easy to change back to civilian vocabulary even after so many years have elapsed.

The Marine Corps train their recruits coming from the East Coast at the Marine Corps Recruit Depot (MCRD) at Parris Island. Those from the West Coast train at the Marine Corps Recruit Depot (MCRD) in San Diego. The training is the same at both instillations but the graduates of both will swear that their training and discipline was much more extreme. Since I graduated Recruit Training from Parris Island I have to agree

INTRODUCTION

with those that swear that Parris Island is much harder. (Just kidding.)

Parris Island is located about five miles from Beaufort, South Carolina and has an area of 8,095 acres. It is reached via a causeway and a bridge over Archer's Creek and is the only way onto the Island except by boat. The military airport on the island was no longer used during my training there and it remains unused today except for recruit training exercises.

The Marine Corps started training recruits on Parris Island in 1915 and they used a ferry to transport the recruits from nearby Port Royal to the Parris Island docks. In 1929 the causeway and bridge were completed, making access to the island much more convenient.

Recruit training in 1966 consisted of three phases. The First Training Phase consisted of learning how to make your bunks, or "racks", how to march and drill, wear your uniform, attend classes, be issued a rifle, take the Initial Strength Test (IST),First Aid, military courtesy and traditions, and more.

The Second Training Phase consisted of moving to the Rifle Range, snapping in, firing for score, serving on mess and maintenance, close order drill, and more.

The Third Training Phase was the final phase in which guard duty was performed, uniforms were fitted and tailored, classes, drill evaluation, run the Physical Fitness Test, (better known as the Physical Readiness Test) and more. This was the phase where you also prepared for graduation.

INTRODUCTION

After boot camp we advanced to Infantry Training Regiment (ITR) which was conducted at Camp Geiger, North Carolina. This was where almost the complete inventory of Marine Corps individual weapons were fired and made familiar to all Marines. Marines with an Infantry Military Occupational Specialty (MOS) received six weeks of training at ITR, while those with a non-infantry MOS received four weeks training.

After ITR, Infantry Marines went on to train in their weapons specialty such as rifleman, mortar man, machine gunner, etc. Non-infantry Marines went on to their assigned school to train in their own specialty. I was given the MOS of 3051 and was assigned to Supply and Administration School to be trained as a Warehouseman.

Altogether it took nearly six months of training to become a fully trained Marine and to receive my first permanent duty station assignment. This book is all about my journey to complete that training, which was more difficult than I could ever have imagined. I was a very reluctant Marine as a draftee but I came to love the camaraderie and pride that bind all Marines together. No book ever written could adequately recreate the atmosphere, intrigue, and physical and mental discipline needed to complete Marine Corps Boot Camp. I hope that my effort to do so will to a small degree come close to projecting some of the unique circumstances every recruit of boot camp endures. Only by being there and surviving it could one fully understand the mystique surrounding Parris Island. I shall always wear the title of **United States Marine** with pride. After all, like the old TV commercial used to say," I earned it."

Chapter One

EXPANDING INTERVENTION

"For all I know, our Navy was shooting at whales out there." With those words President Lyndon Johnson commented on the Gulf of Tonkin incident that would be the pretext to the increased involvement of the United States in the Vietnam War.

On Tuesday evening August 4, 1964, President Lyndon Johnson appeared on TV in living rooms all across America and announced that the U.S. would be launching military retaliatory strikes on the territory of North Vietnam. These attacks would be in response to attacks against two U.S. Navy destroyers (the *Maddox* and the *Turner Joy*) that had occurred on August 2, and August 4, 1964.

The August 4 attack was met with much skepticism by President Johnson, although there was little doubt in his mind that the August 2 attack by North Vietnamese torpedo boats against the two destroyers had occurred. The August 4 attack on the same ships will forever be shrouded in doubt and mystery by many skeptics of the incidents.

Regardless of the reasons for American intervention in an undeclared war between North and South Vietnam, it didn't take Congress long to act on President Johnson's request for authorization for offensive action against North Vietnam. On August 7, Congress enacted a resolution which authorized the President to "repel any armed attack against the forces of the U.S. and to prevent further aggression."

With the passage of this resolution the wheels were set in motion to increase the military forces of the United States to meet the needs of waging an undeclared war, but nevertheless, a war that would require additional human and materiel resources. Resorting to the draft was the quickest and simplest way to do this.

Induction statistics point out that a total of 1,857,304 men were drafted from August 1964 to February 1973, when the draft officially ended.[1] The total draft for the Marine Corps would total 42,633 over the same period.[2] The total number of draftees more than doubled in 1965 from what it had been in 1964.[3]

Little did I know that night of August 4, 1964 while my family was watching President Johnson on TV as he made his speech that just eighteen months later I would be one of those draftees that made up the Marine Corps total of 42,633 draftees of the Vietnam War Era? I had, of course, been aware that the radio and TV news had been dominated by happenings in the Western Pacific nation of Vietnam since May 1961 when President Kennedy had sent 100 Special Forces troops to South Vietnam.

The Tonkin Gulf incident marked a major change in the U.S. involvement in the war and I was aware after President Johnson's speech that August evening that I was in the age group of young men that would be called on to fight this new war, our generation's war, if it wasn't settled quickly. On the

---

[1] http://www.endusmilitarism.org/sssinductionstats.html
[2] http//history-world.org/Vietnam_war_statistics:htm
[3] Ibid.,Vietnam war statistics

one hand I was very saddened by the prospect of the potential loss of lives and treasure that war brings, but at the same time I was excited when I thought of the stories I had read and heard from veterans of other wars about fighting for their country.

That evening as Mom, Dad, my brothers and sisters and I were sitting in front of the TV and watching President Johnson making his speech I turned to Dad and said, "Do you think we'll be in a major war over this incident?" "I don't know, he replied, but sometimes big wars start from little wars." I looked at him as I exclaimed, "Now that I've reached the age of seventeen (the month before) will you or Mom sign for me to go into the service if we get involved in a war?" He stared at me for a moment before replying, "The best thing for you to do is to go back to school this fall instead of thinking about the service!" (I had dropped out of school in the spring of 1964.)

My naïve questions had gone about as far as I thought I should take the subject so I changed the conversation to my impending marriage which was scheduled for Saturday August the 15th. "If you won't sign for me to go into the military can I count on you or Mom to sign for me to get married the 15th of this month?" Again he stared at me for several seconds as he formulated his reply. "I'm not signing anything. If you need anything signed you'll have to ask your Mommy. I think you're too young to think about the military or marriage right now."

I turned to Mom and said," Mom you are going to sign the marriage license for me, aren't you?" She had already promised that she would sign but I needed reassurance just then. She replied, "I said I would and I'll keep my promise, but I could never see myself signing one of my children into the military." I figured I had won the most important of my objectives so I contented myself with their answers for the moment.

I had spent my childhood years in absorbing everything I could about the military, especially enjoying reading about the War Between the States, World War Two, and the Korean War. While most of the neighborhood kids were playing outside in the sunshine during the summer, I spent most of my time in the local Whitesburg library or at home, reading about the exciting Battle of the Bulge, the Normandy Invasion, Tarawa, Gettysburg, and other famous episodes involving our American military forces. It's no wonder that I felt excitement and sadness at the same time when learning that we might be headed for a war that would probably include my generation. My youth protected me from thinking of all the consequences that conflicts can have on lives and families.

My Dad had been drafted in World War Two into the Army and had spent over three years in stateside and overseas service.[1] My Uncle Leonard Sexton had spent at least eight years in the Marine Corps in the 1920's and 1930's and between twenty and twenty five more years in the U.S Army during World War Two and Korea. He had been captured by the Germans during the "big" war and wounded during his service in Korea.[2] Their service had made me very proud of them and motivated me to be very patriotic in my thinking. I grew up thinking that someday, somehow, I would also be a member of the U.S. military.

---

[1] Dad's division was scheduled for the invasion of Japan when the war ended.
[2] Leonard was captured near the end of the war in Germany.

My playmates and I would sometimes slip into Granny's house and into the bedroom closet where she kept Uncle Leonard's U.S. Army uniform hanging where he had left it on one of his infrequent visits. We would admire his ribbons and medals and his E 8 master sergeant's stripes on the arms of the "Ike" jacket of his from World War Two and speculate as to how many stripes our uniforms would have when we joined the military. Not if but when. We were positive it would happen someday. We just didn't know when.

After my conversation with Mom and Dad of August 4, I turned my attention to mine and Wanda's preparation for our upcoming marriage on August 15. I put my 1954 Ford Custom car (Hurst floor shift included) up for sale in order to acquire enough funds to buy the necessary license, to get the required blood test, and for gas money to travel to Clintwood in Dickenson County Virginia in order to tie the knot. I quickly sold the Ford to a friend for $165.00 which gave me enough money for my purpose. We didn't have a place of our own to live, as we had not thought that far ahead. Being young, we just assumed we could go on living with one or both of our parents.

On Saturday, August 15, 1964, I got up at 6am and dressed up as much as I could in my black pants and blue and white checkered shirt that would be my wedding clothes. Dad got up at the same time and left for his job at the Stumbo Supply Lumber Company. He seemed unusually quiet that morning and didn't mention the marriage to be. I realize now that he hated to see me get married so young and he was expressing his feelings by his silence.

Mom was up and getting ready as I went out the door to walk up the one lane dirt road to meet Wanda at her parent's house.

(I either didn't consider or didn't know of the custom of the groom not seeing the bride until the actual ceremony on their wedding day. Besides we had to travel to Virginia together anyway.) When I reached Wanda's home I knocked and went inside to wait on her to get ready to go to Virginia. She didn't seem overly friendly to me that morning and seemed to be deep in thought. I don't know she was just nervous or whether she was having second thoughts at that moment.

Wanda's Mom, Mertie Adams, was also going with us as she would have to sign for Wanda to be married. Wanda and her Mom were soon ready and we went out the door and started down the hill to the car. Winford, Wanda's Dad, didn't have a lot to say that morning, but as we went out the door he cautioned us to "Be careful."

Wanda looked beautiful in her wedding outfit which consisted of a red shell sleeveless top and a brown plaid skirt. Her Mom and mine were also dressed in nice skirts and blouses for the wedding that we were looking forward to take place that day.

When we reached the bottom of the hill where Dad's 54 Chevrolet Power Glide was parked we noticed my Granny sitting on her son Denver's porch. She was there to see us off but was feeling very melancholy about us two youngsters getting married and was having a difficult time hiding her feelings. (She may have been experiencing a foreboding of bad news she would receive later that evening.)

When we arrived at Clintwood, Virginia about one hour and fifteen minutes later we proceeded to the Dickenson County Courthouse and applied for our marriage license. Mom signed the application for me and Mertie signed for Wanda. We then went down the street for the blood test and waited an hour or so for the results.

When the blood test results came in we went a little further down the street where the Reverend Marion A. Coe performed the marriage ceremony. Regretfully, no portraits of the ceremony were taken that day. My thinking hadn't extended very far on that subject either. The only others present for the ceremony besides me, Wanda, and the Reverend were our Moms who once again were our witnesses.

The trip, the application process, the blood tests, and the marriage ceremony had been completed before the arrival of the noon hour on that Saturday, August 15, 1964. I don't think we as a couple have ever missed the complications and hassle of a formal wedding. Somehow the simplicity of an informal wedding proved to be as memorable, and in some ways much better than a more dignified church wedding would have been.

We returned to Tunnel Hill and pulled into Uncle Denver's yard just after the noon hour. Granny was still sitting in the same spot we had left her in. I'm sure she had done a lot of worrying while we were gone but outwardly she seemed happy for us as we recounted the day's events to her.

Shortly after our arrival and while talking to Granny, a car pulled up containing Uncle Denver and Dr. Carl Pigman. As they got out of the car and walked up the steps to the porch where we were all gathered it was obvious by the look on their faces that something was wrong.

Dr. Pigman pretended he was there for Granny's routine checkup. (He sometimes made house calls for elderly people.) As he finished checking her pulse and heartbeat he said to her,

"Mrs. Sexton is you feeling alright today?" She looked at him as if she sensed he was there for another reason. She then replied, "Yes Dr. Pigman, I'm feeling as good as someone my age can be expected to feel." He then said, "I'm afraid I have some bad news. Your son Leonard passed away with a heart attack this morning." Without replying, the tears started streaming down her face as she reached for her handkerchief she always carried in her purse. As she dabbed at her eyes she said, "I knew something was wrong when I saw you and Denver pull in the yard together. I have had a dread on me all day."

The news of my Mom's oldest brother's passing had my Mom sobbing softly as she stood there on the porch and Uncle Denver had tears in his eyes. At that moment I was pulled between my happiness in getting married and the sorrow of my Uncle Leonard's death. With his death one of my military heroes was no more.

After a few weeks of marriage Wanda started working at Pigman Brothers Dry Cleaners. We were staying with my parents and the crowded conditions in their small home caused us to want to hurry and get our own place as soon as we could.

I got a job with Ramsey Well Drilling which was located at the foot of Tunnel Hill on the Pine Mountain Junction side. I worked there during the winter months which made for some very cold days, especially in the early mornings. The cold steel bits we had to change often while drilling would freeze your skin to the steel if you weren't careful. I lost plenty of hide on that job.

The pay was very good and we saved a little bit of money to enable us to buy some furniture and begin looking for a house of our own. Eventually the small frame house located just below

Dad's and Mom's house became available for rent. Wanda's Aunt Opal was then living there and had decided to move elsewhere. We rented it for $25.00 a month and sometimes struggled to pay the rent. It didn't take us very long to move our meager belongings the 25 yards to our new residence.

After four months with Ramsey's Well Drilling I took a job with Kyva Motor Company in Whitesburg, Kentucky. This company was the local dealer of Buick and Pontiac cars. My job consisted of helping with the preparation of the new vehicles for sale, helping service them, and preparing used cars for sale. My immediate co-worker was Hiram Stallard of Whitco, Kentucky.

In January or February, 1965, an opening for an assistant manager/clerk position at our local grocery market became available after the person holding that position took another job. J. Don Collins was the owner of the supermarket/convenience store which was located within sight of our house. The decision to change jobs was an easy one to make.

The pay in my new job was $.80 an hour and after a short time I got a nickel an hour raise. The pay was pretty fair, especially since I was getting 74 hours a week in my seven day a week job. I enjoyed the responsibility of doing most of the ordering and stocking of grocery items and I also enjoyed meeting and greeting our many customers each day.

Meanwhile the war news heard on the radio and TV was beginning to escalate with each passing day it seemed.

Chapter Two

MILITARY OPTIONS

Finally reaching the age at which I could legally make decisions for myself, including the right to sign my own enlistment papers if I wished, I began to think seriously about my options concerning the military draft. At that point in time of our nation's history every young man of draft age had to decide if he wanted to go to college, try to find a steady good paying job where an understanding company would hire him despite his likely draft status, or if he wanted to go ahead and perform his military obligation by joining a branch of the military of his choosing.

My own options were considerably limited by the fact that I was already married and my first obligation was to my wife. She and I discussed the matter many times in the year of 1965 and I was always burdened with wanting to get my obligation over with so I wouldn't always have the threat of being drafted in the background of any plans we might have for the future.

After much discussion and plenty of worry, my wife finally gave her approval to fully explore my options with the different branches of the military. Whatever I chose to do or not do would be fine with her. She accepted the fact that if we didn't meet the problem head on we would always have the possibility of my being called for duty until I reached the age of twenty six, which was the last year of eligibility for the draft. Like most young people we were much too impatient to wait that long to plan for the future.

After receiving Wanda's approval to actively begin to seek a military option that would best fit our needs, I felt some relief because of the pressure I had been feeling about maybe having to make a decision without her approval. She had agreed with me that planning for any kind of a future was difficult enough without the possibility of having to stop housekeeping and to start over again if I was drafted. I had already researched the starting pay for an E-1 private in the military and the result wasn't encouraging. Keeping up a house on $87.90 a month with all the ensuing bills would be almost impossible.[1] These were difficult facts and decisions for a couple of teenagers to have to make but it helped us to grow up quickly, that's for sure.

I was still working at Hilltop Grocery and had saved enough money to purchase a 1956 Ford Crown Victoria automobile in the summer of 1965. I had always wanted a Ford C. V. but the car turned out to be an unreliable mode of transportation to say the least. It was good enough to drive locally but wasn't to be trusted on a long trip. I eventually sold it for what I had paid for it, a cool $100.00.It seems as though I either sold all my vehicles at a loss or for what I had paid to purchase them. Making a profit wasn't in my repertoire I suppose.

---

[1] After a couple of months a spouse would receive $60.00 per month dependent allowance

The loss of out mode of transportation wasn't all that inconvenient for us as my workplace was less than five hundred feet from our home which was located across the road from Stumbo Supply on Tunnel Hill. Any time I needed to make a trip to town I could always depend on Dad letting me borrow his car.

In the fall of 1965 I decided to try for my General Educational Development (GED) Diploma. I had dropped out of high school in the spring of 1964 and it was pretty obvious to me by then that I needed my high school diploma to have any measure of success in life. Besides, I wanted to soon begin my military branch search and I wanted to make sure I had some extra tools for a successful vocation in life, whether it involved a military career or a civilian occupation.

I contacted the Letcher County Superintendent's office to schedule a date to take the preliminary test to see if I was prepared to take the GED test. I passed that first hurdle to a diploma and the Superintendent's office obtained an appointment at Pikeville College in Pikeville, Kentucky for me to take the necessary tests. The testing procedure required you to take one test per day which meant I would have to take the five separate tests over a five day period.

My boss, J. Don Collins, offered to let me drive his new Ford automobile to travel to Pikeville, and allowed me to take mornings off the week of the scheduled tests. He took an interest in wanting me to succeed in life and went out of his way to help me in any way he could.

After I had completed the five tests I had to wait a week or two until the Letcher County Superintendent's office called me to inform me that I had successfully passed all the tests to obtain my GED. They said the diploma would be sent to me shortly. As it turned out, I would have a long wait for that diploma.[1]

A few days after taking the GED tests I was behind the counter at Hilltop Grocery when I watched as a car came into the store parking lot from the highway and crashed into the pop machine located just outside the store window where I was standing. The impact knocked the pop cooler completely over but it had kept the car from coming into the store where I was standing right beside the window.

The close call rattled me a little but I rushed outside to find a local well known man climbing from the banged up car. He turned out to not have a scratch on him but was drunk as a skunk and didn't know what had happened or where he was.

Seeing that he was ok I rushed back inside the store and called J. Don and explained what had happened. I then asked if I should call the law but he said, "No, just wait. I'll be right up and check it out." He was there in less than five minutes and after assessing the damage he decided to just let it go, as he knew the man and his family well. The drunk had finally jumped back into his damaged car and skedaddled just before J Don arrived there. J Don took the incident much better than most people would have.

My close call at the store convinced me that one could get injured or killed at work as quickly as in the military sometimes. I had seen a few serious accidents while working part time at

---

[1] I received my diploma exactly two years later in November, 1967 while stationed at Camp Lejeune, North Carolina

the Stumbo Supply Lumber Company but this was the nearest I had come to being seriously injured in any of my work places.

I had to be at work at nine o'clock each morning and I got in a habit of setting my alarm clock for seven o'clock each morning to catch the CBS newscast on our local radio station before coming to work.[1] I would listen as the announcer rattled off the name of a military operation that our troops would be involved in against the Viet Cong or against the North Vietnamese regular troops. I was listening the morning that CBS announced that they would no longer be giving individual casualty statistics but would instead give the U.S. casualties as being light, medium, or heavy. I mentioned to Wanda that morning that things must be getting pretty hot over there if the casualty count was going to be kept hidden from the public.

Along with the news about the military operations and the casualty reports from Vietnam, the news would sometimes mention the number of draftees expected to be called from month to month. By the end of 1965 the total number of draftees for that year was 230,991. The number for the preceding year of 1964 was 112,386.[2] I always paid close attention to those numbers in those days.

My friend from my old Kyva Motor job, Hiram Stallard and I had been discussing trying to get into the Air Force for the past few weeks and we had agreed that we would go to the Hazard, Kentucky recruiting station to see what that branch of service could offer us in the way of some type of technical education opportunities. We both were interested in some type of job that might benefit us in the civilian world as well as the military. We

---

[1] We didn't have a TV set.
[2] http://www.endusmilitarism.org/sssinductionstats.html

both were leaning towards trying to get into aircraft engine mechanics or hydraulics if we could qualify for those jobs.

One cold morning in November, 1965, we climbed into my Dad's old Chevy and drove to the Hazard Recruiting Station. We didn't have an appointment but luckily the recruiter was in and wasn't busy at the moment.

After we had discussed our options with the recruiter he informed us that there was a waiting list to join the Air Force but we could still take the qualification tests if we wished to. He said that even if we passed all the tests it could be several months before we would be able to receive our basic training at Lackland Air Force Base in Texas. We decided to take the tests anyway and see what happened. The recruiter then scheduled us to take the ASVAB or whatever the test was called at the time.

Approximately two weeks later we took the required test or tests and waited for the results. The recruiter called us back by phone in a few weeks to inform us we had passed the first part of the process and we needed to come in for some more paperwork and tests. So we went back to Hazard and completed some more paperwork and went home to wait some more. This was becoming more like buying a house than joining the military, I thought. (Not that I had ever bought a house.)

I had another good friend that had grown up with me on Tunnel Hill who had expressed an interest in joining the military with Hiram and me but he, Danny Reed, didn't show any interest in joining the Air Force with us. He was leaning more towards the Army and wanted to pursue that path first. I told

Danny that if I changed my mind I would check out the Army next.

As November turned into December, 1965, I decided to have a backup plan in case my Air Force deal didn't happen quickly enough for me. I have always had a tendency to be impetuous at times and this occasion was one of those times. I decided to check into the Selective Service system to see where I stood with the Draft Board and the draft. Since I had been prequalified 1-A when I registered with Selective Service on my eighteenth birthday, I wanted to know my likely draft date if possible.

I spoke with my friend Danny Reed about checking on my draft status and he asked if he could tag along and ask about his status also. We went to Whitesburg the first week in December and walked to the U.S. Post Office. We went into the basement where the draft board was located and posed the questions of when we might be expected to be included in the draft and what were the chances of our not being called?

The clerks there either couldn't or wouldn't tell us when or if we might get our call but they did say that there was one way of being sure of being included in a future draft. Since us both wanted badly to get our service time over and done with we were quick to ask how that was possible. The clerk explained that we could sign a paper to be included in the next draft call or soon as possible thereafter.

Danny and I decided to take a walk through town before we committed to that giant step. This was something we hadn't considered doing but since we were anxious to get our military time over with and get along with our lives this quick path to the draft might be just what the doctor ordered. After standing on the street and discussing it for a good while my impetuous nature got the best of me. I said to Danny, "I've already, for all intents and purposes, joined the Air Force and I'm just waiting for their call. I've been thinking about the Air Force and the required four year enlistment and trying to weigh that against the two years the Army requires of draftees." Danny looked at me and said, "Eddie, two years sounds much better than four, especially if you should get in the service and not like it." I had been thinking along those lines myself and those words sounded logical to me.

After discussing the situation we crossed the street to the post office and once again headed down the basement steps located beside the post office and then opened the door to the office of the draft board. I felt a little apprehensive about signing papers that would subject my choice of jobs in the military to the whims of a bored clerk in a military personnel office. At the same time I felt relieved that I could now set aside the agony of not knowing how to plan for the future, at least for the moment.

Signing the papers for the draft was much easier and simpler than testing and signing various papers for the Air Force, I thought. As that thought ran through my mind I wondered what I would tell the Air Force recruiter now that I was sure to be included in the draft in the near future. I now had a decision to make whether to take the Air Force route if the

recruiter called back before the next draft or whether to wait on the draft to call me. At least now I had two options that I could control somewhat. It wasn't like the dilemma I faced before when all my plans for a future were on hold to see what Uncle Sam had in store for me.

As Danny Reed and I left the post office and headed home I had another decision to make. What would I tell Wanda about the visit to the draft board that morning and the steps I had taken to settle the matter once and for all? I then made a very poor choice. I decided not to mention the trip at all. Wanda was expecting me to join the Air Force, as I had told her everything about the testing and signing episodes, but I hadn't mentioned to her about my plan to check with the draft board. I justified my thinking by telling myself that she had already given her ok with checking the Air Force out so I assumed the choices weren't that far apart.

Regardless which branch of the military called me first I knew one thing for certain. Wanda and I would soon have to be apart for awhile during any basic training I would have to take. After that, who knows where I would be sent?

I continued working at Hilltop Grocery throughout the month of December while waiting for the Air Force to call or to receive the notice from the draft board. I told J. Don that I would most likely be leaving within a couple of months at the latest.

When the month of January rolled around I expected every day to hear something from the Air Force recruiter or the local # 58 draft board, but the first two weeks dragged by without a word from either of them. I was becoming very impatient and wanted to hear something, and soon.

Wanda had started working at Hilltop Grocery part time when she could to help out with the bills. She still wasn't aware that I might be in line to be called by the draft but she wondered with me why I hadn't heard from the Air Force yet.

We still hadn't decided where Wanda would live or where we would store our few pieces of furniture if and when I left for training. I guess we just didn't want to think about it for any length of time even if we knew it was inevitably going to happen.

I hadn't told my Granny, Nancy (Coonie) Sexton, about my potentially leaving soon for the military and I dreaded breaking the news to her. Wanda and I walked up the hill Saturday January 22, and finally told her. As expected she didn't take it too well and cried softly and stayed silent for several minutes. She said she had worried about her son Leonard so much when he went through two wars that she dreaded to see any of her grandchildren have to go through that. I tried to reassure her but I don't think I changed her mind at all. As we left her home I told Wanda that I regretted that my decisions were hurting so many family members but I couldn't see any other way out of our uncertain situation.

Chapter Three

FINAL EXAMS

The Waiting is Over

Figure 1 The Former Whitesburg Post Office

After my visit with Granny I felt better knowing that all of my family now knew that Wanda and I were just waiting for the call that could come at any time. It was inevitable that I would feel a touch of depression and gloom at the prospect of having to leave our home after just now being able to get on our feet (financially speaking) because of Wanda's part time job at the supermarket. It wasn't as if we had a cornucopia of new found wealth but at least we were now able to pay our bills in a timely manner.

I had convinced J Don to sponsor a bowling team that I organized for a bowling league that met every Tuesday night at Whitesburg, Kentucky. I didn't start the team just to allow Wanda to get a few hours of work but that's the way it turned out. I was going to miss our bowling nights.

On Friday, January 28, 1966, I went to the post office in Whitesburg to check our mail. I was nervous and fumbling as I opened box # 215 as I knew the letter from the draft board could come at any time. We usually only checked our mail once or twice a week because of the inconvenience of having to borrow Dad's car and trying to find a parking place in town. As I pulled the mail out of the box I saw the word IMPORTANT written across a white envelope and right away my heart jumped.

I knew that this was the letter I had been dreading to receive ever since Danny Reed and I had made a trip to this very post office building the first week of December to check our draft status. I didn't dread it for myself so much as I did for my wife and the unknown change to our unsettled circumstances this letter would invoke. I still hadn't told her all the details about my visit to the draft board in December. She had no idea this letter was coming, even though she had understood the possibility existed of my receiving one eventually.

I hastily examined the letter and sure enough in the left hand corner the words "Selective Service" were prominently displayed in large letters. There was no turning back now. This was it. If the Air Force recruiter didn't call soon and give me a choice I would be going into the Army as a draftee now.

As I left the post office and walked to the parking area on the riverside road near the Salyers building, I felt almost numb as I contemplated the difficulty I faced in telling Wanda about the letter I had received. I hadn't really given much thought to that part of the equation when struggling to make a decision about what to do about my military obligation. Like a light had just switched on inside my head I realized then that most of my decisions had been of a selfish nature while pondering my direction in life. Too many of my plans had concentrated on my wishes and desires instead of *our* needs and wishes.

Those thoughts made my dread of telling Wanda even more painful, if possible. Those two miles between Whitesburg and Tunnel Hill seemed to be so many more miles than usual on the drive home that day. I still cringe when I think of how sad I felt to the core of my being while going back home that early Friday morning with that unopened letter lying beside me in the front seat of Dad's '54' Chevrolet.

Arriving home I parked Dad's car in his driveway and walked the few yards to my house and went inside to find Wanda busily cleaning house. She seemed to never stop doing cleaning even if she was going over something she had already cleaned that day. Even simply observing her as she cleaned the house made me dread more than ever telling her about the letter. I remembered how proud she had been when we managed to afford renting our own place just a few months ago. She scrubbed and cleaned our $25.00 a month home like we had a mansion instead of a plain weather worn old four room house. Those thoughts and feelings made my task even harder for me than I had anticipated while driving home.

I asked her to sit at the kitchen table with me to discuss something important. After we had sat down I pulled the letter from my jacket pocket and showed her the front of the letter as I pointed to the return address in the left hand corner. I saw her eyes visibly moisten but she never spoke a word as I opened the letter and read the words to myself "Order to Report for Armed Forces Physical Examination." I looked over the page which stated that I was to report to the Post Office in Whitesburg, Kentucky on "9 February, 1966 at 5am EST." I then handed her the letter without comment.

As she finished her reading and handed me the letter back, she said in a somewhat subdued voice, "Does this mean we'll have to give up our house and quit housekeeping?" Of course I didn't know myself but I wanted to give her some hope as I replied, "Maybe not. If I pass the physicals and have to go to basic training you'll be able to go with me if I'm stationed in the states." "That's a big maybe, she said. What if they send you overseas?" I had no good answer for that question but assured her that whatever happened I would try to see that we were together as soon as possible.

I went to work that morning carrying my letter to show my boss, J Don Collins. None of us knew yet if I would even pass the physical that I was being called to take but we had to plan as if I would be leaving. J Don said he would hire someone to begin training to take my place but he would make it clear to them that the promise of a job hinged on my going to the military. If for some reason I didn't go, the job would still be mine. He would also still employ Wanda part time if she wished.

Almost unbelievably a phone call came that same Friday afternoon from the Air Force recruiter to my Dad's phone. Mom answered the phone and took the message. She told the recruiter I would be near the phone after 9 pm if he would call back. Mom then called me at the store to inform me of the call. I couldn't believe both calls to me had come on the same day.

After receiving the call from Mom I had difficulty keeping my mind on the business at hand as I stood behind the counter and waited on customers. I tried to mull over in my mind whether to accept the call from Uncle Sam or if I could legally still join the Air Force if I had been accepted. I was torn between choosing the draft and just getting on with my life or possibly waiting several months for an opening in the Air Force. Everything would hinge on what I heard from the recruiter later that evening.

SELECTIVE SERVICE SYSTEM   Approval Not Required.

ORDER TO REPORT FOR
ARMED FORCES PHYSICAL EXAMINATION

Local Board No. 58
Selective Service
Post Office Building
Whitesburg, Kentucky 41858

To   Eddie Gregory Nickels
        P. O. Box No. 215   (Local Board Stamp)
        Whitesburg, Kentucky 41858
                                       24 January 1966
                                       (Date of mailing)

SELECTIVE SERVICE NO.

15	58	47	215

You are hereby directed to present yourself for Armed Forces Physical Examination to the Local Board named above by reporting at:

Local Board No. 58, Rm. No. 1, U. S. Post Office Bldg, Whitesburg, Ky.
                                    (Place of reporting)

on   9 February 1966   at   5 AM EST
      (Date)   (Hour)

                                    Lelia H. Banks
                                    (Member or clerk of Local Board)

IMPORTANT NOTICE
(Read Each Paragraph Carefully)

TO ALL REGISTRANTS:

When you report pursuant to this order you will be forwarded to an Armed Forces Examining Station where it will be determined whether you are qualified for military service under current standards. Upon completion of your examination, you will be returned to the place of reporting designated above. It is possible that you may be retained at the Examining Station for more than 1 day for the purpose of further testing or for medical consultation. You will be furnished transportation, and meals and lodging when necessary, from the place of reporting designated above to the Examining Station and return. Following your examination your local board will mail you a statement issued by the commanding officer of the station showing whether you are qualified for military service under current standards.

If you are employed, you should inform your employer of this order and that the examination is merely to determine whether you are qualified for military service. To protect your right to return to your job, you must report for work as soon as possible after the completion of your examination. You may jeopardize your reemployment rights if you do not report for work at the beginning of your next regularly scheduled working period after you have returned to your place of employment.

IF YOU HAVE HAD PREVIOUS MILITARY SERVICE, OR ARE NOW A MEMBER OF THE NATIONAL GUARD OR A RESERVE COMPONENT OF THE ARMED FORCES, BRING EVIDENCE WITH YOU. IF YOU WEAR GLASSES, BRING THEM. IF MARRIED, BRING PROOF OF YOUR MARRIAGE. IF YOU HAVE ANY PHYSICAL OR MENTAL CONDITION WHICH, IN YOUR OPINION, MAY DISQUALIFY YOU FOR SERVICE IN THE ARMED FORCES, BRING A PHYSICIAN'S CERTIFICATE DESCRIBING THAT CONDITION, IF NOT ALREADY FURNISHED TO YOUR LOCAL BOARD.

If you are so far from your own Local Board that reporting in compliance with this Order will be a hardship and you desire to report to the Local Board in the area in which you are now located, take this Order and go immediately to that Local Board and make written request for transfer for examination.

TO CLASS I-A AND I-A-O REGISTRANTS:

If you fail to report for examination as directed, you may be declared delinquent and ordered to report for induction into the Armed Forces. You will also be subject to fine and imprisonment under the provisions of the Universal Military Training and Service Act, as amended.

TO CLASS I-O REGISTRANTS:

This examination is given for the purpose of determining whether you are qualified for military service. If you are found qualified, you will be available, in lieu of induction, to be ordered to perform civilian work contributing to the maintenance of the national health, safety or interest. If you fail to report for or to submit to this examination, you will be subject to be ordered to perform civilian work in the same manner as if you had taken the examination and had been found qualified for military service.

SSS Form 223 (Revised 4-28-65) (Previous printings may be used until exhausted.)   ☆U.S. GOVERNMENT PRINTING OFFICE: 1965—O-774-905

Figure 2 This is the original letter I received from the draft board on January, 28, 1966

After J Don and I closed the store that evening, I crossed highway 15 and proceeded to my house where Wanda was waiting to go with me to Mom's and Dad's to await the phone call from the recruiter. She was as anxious as I was to see what he said. I still had no idea what my decision would be if the recruiter was calling to inform me of my acceptance into the Air Force. For all I knew he was going to tell me I wasn't qualified after all the testing I had gone through, which would be a disappointment regardless of the choice I would make.

We walked the short distance to Dad's house and sat down and discussed the decision I soon might have to make. We were still debating what would be my wisest choice when the phone rang. After a few words of pleasantry conversation we got down to business.

Recruiter; "Eddie, I've called with some good news for you. You've been approved for enlistment into the United States Air Force on the 'buddy plan' with Hiram Stallard and I have an appointment date for your basic training at Lackland Air Force Base in Texas. You just need to come in and sign a few more papers to make everything official." As I listened to what he was saying I was still trying to make up my mind. When he mentioned at the end of his talk that I needed to "come in and sign a few more papers" I immediately made up my impulsive mind to go with the draft for two years instead of the Air Force for four years.

I steeled myself as I answered his statement with these words, "I appreciate all you've done for me Sergeant, but I've just received my Selective Service letter to report for a physical for pre-induction and I really want to explore that option first. I've given it a lot of thought and I believe I'll just take my two year obligation over the four years the Air Force would require me to serve." He replied, "Are you certain you want to give up this slot after all the waiting you have gone through?" "I'm certain that I'm making the right choice to suit my circumstances, I replied. I feel that after serving two years I'll know for certain if I want the military for a career or if I want to work in the civilian world."

The sergeant than thanked me for my interest in applying for the Air Force and I thanked him for all his help and interest in me. I didn't mention to him that if he had called a few weeks sooner I would have doubtless chosen a different pathway but that explanation wouldn't have made a whit of difference in the events of that evening. I held my tongue. He urged me to call him if I changed my mind about taking my chances with the draft. I assured him I would but I pretty much had my mind made up. With those words I hung up the phone and my destiny was sealed.

With less than two weeks remaining until I had to go for my pre-induction exam, Wanda and I began trying to make plans as to whether she would continue to rent our home when and if I was determined to be physically fit for the military. We could never really come to a conclusion as to what we should do, so in the end we decided to put any decisions off until after my pre-induction examination. We knew that we would have a month or more after the first exam before any actual draft call might come. We would wait and make the hard decisions later.

The next day I talked to Hiram Stallard on the phone and told him of my decision to not go into the Air Force. He said he completely understood and that he had heard from the recruiter Friday also and that he was going to go ahead and sign up the coming week. So at last that was the end of my life long desire to join the Air Force.

I saw Danny Reed that same evening when I was working at Hilltop Grocery and he told me he had gotten the same letter for the pre- induction physical as I had. He wasn't married at the time, so he was excited about the possibility of getting into the Army and maybe making it a career. We discussed our letters for a few minutes before he said, "Eddie what kind of a job would you like to do in the Army?" I grinned as I said, "I'm not real particular about my choice of jobs but I'd like to be stationed in the states for as long as I can. After all, I have a wife to take care of." He then said, "Well I only have myself to take care of so I don't care where they send me. As a matter of fact I intend to volunteer for the paratroopers if I get a chance." I told him I had no desire for the paratroopers as I had tried and failed at that when I was a young boy! (I had previously told him about the time I jumped off our high side porch with Mom's umbrella and knocked the breath out of me as I rolled several yards down the hill into our corn patch.) I had no desire to repeat that episode in my life.

The next few days went by quickly and on Tuesday February 8, 1966, I worked my regular shift at Hilltop Grocery. I got home a few minutes after 9pm and went straight to bed to try to get a little sleep before I had to get up at 4am. It was no use, as I didn't sleep a wink that night. I had always had a tendency to be sleepless when planning something exciting for the next day.

Going to be examined for the military was scary and exciting at the same time in my mind.

I got dressed and slipped out the door without waking Wanda and walked to Dad's front yard to use his car to drive to Whitesburg. It was very foggy that morning and as I drove to town I had trouble seeing the road even though I was familiar with every curve and straight stretch between Tunnel Hill and Whitesburg.

I parked the car just above the Salyers building on the road running beside the Kentucky River. I left the car unlocked and hid the key in the car as Dad had instructed.[1] I planned to drive the car home when we returned to Whitesburg that evening but I left the key in the car anyway in case Dad might need the car for something.

I reported in to the local board in the post office basement and was told the bus would be a few minutes late due to the foggy conditions. The twenty or so potential draftees gathered there that morning were all talking amongst each other and joking about how they hoped the bus would never be there to get them. I agreed with them although I believe all of us were really very anxious to get on the road and get it over with.

Around 5:30 am the bus finally pulled up and parked in front of the post office. The driver opened the door directly in front of the sidewalk that led into the basement of the post office and told us to "load up." After all of us entered the bus and got seated someone came out of the office of the draft board and called the roll. I imagine they thought a few young men might

---

[1] I wouldn't take a chance in doing that in this day and time.

31

slip off home after checking in. It had probably happened before, I suppose.

The bus that picked us up was an older model with a diesel engine. The diesel fumes were spewing throughout the bus, especially going up the hills. The bus was so slow when climbing a hill that the fumes outran the bus and filled the bus until we got a little speed up while going down the other side. I thought sure I would die of diesel fumes before we ever got near Ashland, Kentucky where our examinations would occur.

We traveled through Hindman, Martin, Prestonsburg, and Paintsville on our way to Ashland that morning. Most of the roads were two lane narrow roads which made for a slow and tedious journey to Ashland in those days. There was also no restroom aboard the bus so we made a restroom stop somewhere near Prestonsburg on the way there.

Most of us were so sick on diesel fumes that we talked very little on the way to Ashland. Opening the windows helped a little but it was so cold the windows could only be kept open for a short stretch at a time. We were very pleased when we reached the city of Ashland around 10:30 that morning.

The examinations were held in the Armed Forces Examining Station in the basement area of an Ashland hotel building. (I have forgotten the name of the hotel.) It seemed that there were more than a hundred young men taking physicals and written examinations that day. We were first given our physicals which included every manner of poking and probing procedure you could think of. Most of our exams were done while completely unclothed. If one had any inhibitions upon arriving there I can assure you they didn't have many left when they left that building. Nothing was left unseen or not probed by the examining physicians.

After the physicals we were seated in a large room and asked if we could read and write. A surprising number of young men held up their hands and said they couldn't do either. The speaker at the podium then informed everyone that it was against the law to lie about the ability to read and write just to get out of the service. The question was again asked and only two or three hands remained in the air. I was a little surprised that so many had at first thought they could get by with a lie of that magnitude.

We were fed a nice hot dinner in the hotel dining room and after that we continued taking a battery of written tests. We were reminded to do our very best so as to assure us being assigned a good job or technical skill in the military. I think this information changed a lot of minds for those who had thoughts of underachieving on the tests to lessen their chances of being drafted. I tried my best to score as high as I could on all the tests.

At around 5pm we climbed back on the bus and headed home. The trip back was just as bad as the trip up had been. The diesel fumes made us all sick and we were wonderfully glad to pull back into Whitesburg around 10:30pm that evening. I walked to where I had parked the car and found the key where I had left it, started the car, and drove home to tell Wanda all about my trip. I slept very well that night.

The next day I was back at work and everyone I met wanted to know how the examination went. I told them all about our problems with diesel fumes and that we were told that we would be notified by mail if we passed all the exams .We would probably be informed within a few days of our acceptance or rejection.

Wanda and I didn't discuss our possible separation very much at all after my return from the pre-induction physical. We tried our best to not think of any sad good byes. We knew they would likely happen soon enough anyway.

A few days later I was offered a job by a top tobacco company representative that I had come to know while working at Hilltop Grocery. The job as Product Representative for the company consisted of going to grocery stores, supermarkets, service stations, and other retail outlets and setting up tobacco displays to advertise their products.

Having grown up with a tremendous aversion to cigarettes and the smoke they emitted which usually choked me half to death when I was around them, it wasn't a surprise to my friends when I turned the job down. Some questioned my sanity of turning down a job that paid almost three times what I was then earning but I never seriously considered taking the job. I surely would have been labeled a hypocrite by those who knew me best if I had accepted a job selling a product that I despised.

If I had been offered a job in a different vocation it wouldn't have made any difference. I couldn't have taken it anyway. I was waiting for a date with Uncle Sam.

On Tuesday February 15, 1966 I received a letter entitled, "Statement of Acceptance" from AFES, Ashland Kentucky.

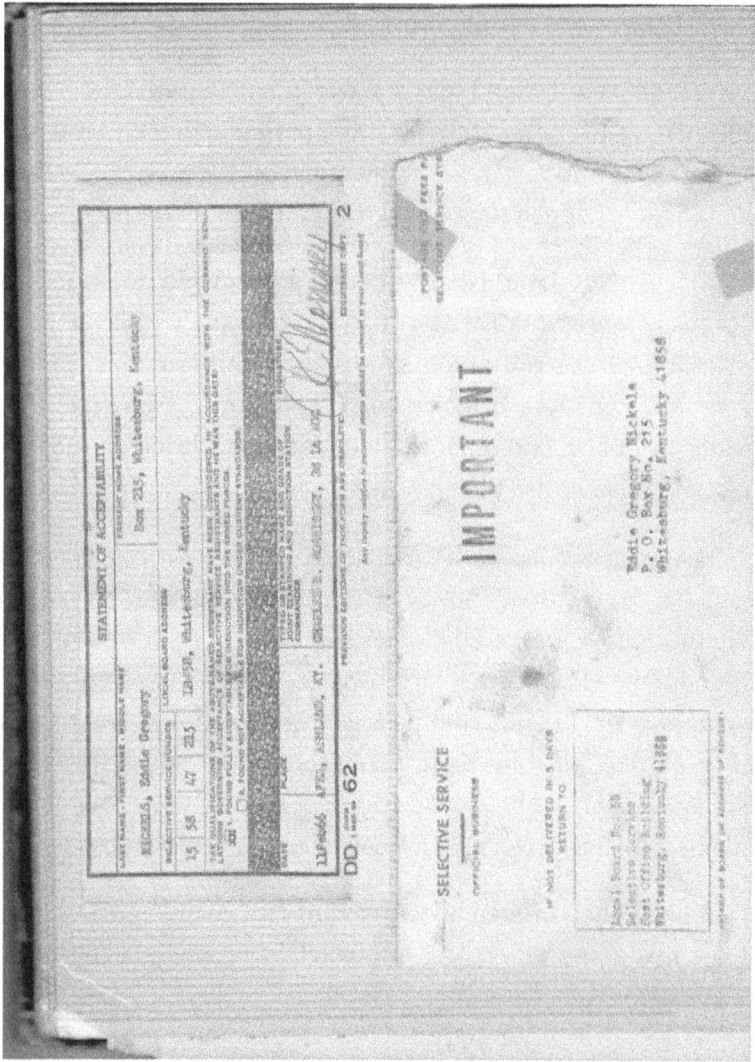

Figure 3 This is the original letter received from AFES, Ashland, Kentucky on February 15, 1966.

Chapter Four

THE CALL TO DUTY

Leaving Home is Hard to Do

Figure 4 Wanda and the author at the time of his induction into the military.

After receiving the letter containing the Statement of Acceptance, Wanda and I resigned ourselves to the fact that we would soon be separated at least for awhile for the first time in the year and a half that we had been married.

Since I now knew that my 1-A draft status was confirmed I notified my boss at work (J Don) that I would almost certainly be in the next draft or the one following. He was still training someone for my job and Wanda was still working part time for him. She would relieve Don and myself out for lunch each working day. She was working at least two hours daily and sometimes more. The extra money came in handy and we saved a little change now and then.

Wanda and I tried to discuss what she would do when I left but our discussions usually brought on such sadness on both our parts that we avoided the subject as much as possible. We finally agreed that we would wait until the second and final letter came from the draft board to make a decision. That gave us a few days of rest from the painful subject.

A day or two after I received my Statement of Acceptability in the mail, Danny Reed came by the store where I was working and said he had also received a letter but his statement was different from the one I had received. He said that his had the box checked that said, "Found not acceptable for induction under current standards." No other explanation was included. He was very disappointed but said that he intended to try to join the Army again at a later time and hoped he would pass all the exams the next time around.

I had half way expected that I might be rejected for service due to a couple of incidents that occurred at the examining station involving my physical exam. As our long line of naked potential inductees was entering one by one into a room to have our blood pressure and heart rates checked, my blood pressure was found to be very high. The doctor conducting the checks took my blood pressure two or three times before asking me, "Young man, have you ever had a high blood pressure problem before?" I replied, "No sir, not to my knowledge." He checked it again and motioned towards a cot in the back of the room as he said, "I need you to lie down on that cot for awhile to see if your blood pressure will come down a little."

I laid on the cot as ordered for about fifteen or twenty minutes as the men kept coming into the room and the doctor and the Army male nurse checked their blood pressures. The picture of me lying on that cot completely without clothing and staring at the ceiling as the men trooped in and out of the room is not a pleasant memory. I imagine they thought I had passed out while taking my blood pressure. I hoped they couldn't remember my face!

After a few minutes the doctor then called me forward and rechecked my blood pressure again. Without comment he told me to go on to the next check station. As I left I thought he said, "They'll check your urine there." I thought that statement was strange as they had taken urine samples early in the day and it was now evening.

I rejoined the line and after a few minutes of waiting I reached a large room where a booth with a glass window was set up in the back of the room. Inside were two soldiers with earphones who were looking into the room where our line of inductees were standing. I thought, this sure is an unusual way to check our urine. Are those soldiers there to watch us? If so, what role do the earphones play in taking urine samples? I was totally confused.

A loudspeaker was hanging on the wall and a voice of one of the soldiers in the booth instructed one of our number to pick up a pair of earphones from the desk sitting there and listen for the sounds coming into the earphones and indicate by a hand held "clicker" which ear the sound was being heard in. It finally dawned on me that the "urine test" was in fact a hearing test instead!

When my turn came to be tested I found I could barely hear the sounds coming into my right ear and I was clicking the hand held clicker only for the left ear which I could hear very well. I finally became so confused that I was just clicking away without really knowing where the sound was coming from. One of the soldiers in the booth said over the loudspeaker, "We know exactly what you're trying to do and it's not going to work. We've had that trick pulled before!" It was obvious that they thought I was faking the hearing test but I had no intention of doing that.

I was never informed at the examining station whether the blood pressure problem or the hearing test might cause my rejection for service. I was therefore somewhat surprised that they had accepted me. The draft was expanding each month during this period so some slight hearing and blood pressure problem wasn't going to be the deciding factor, it seemed.[1]

I still listened to the news at 8 am each morning to catch the latest war news before going to work. On January 29, 1966, the U.S. began bombing around the Hanoi and Haiphong area. This was a major escalation of the war and it was pretty obvious that increased manpower and resources would be needed because of our greater involvement. With my pending call to duty, both

---

[1] Number of total inductions in 1965 was 230,991. For 1966 the number was 382,010
http://www.sss.gov/induct.htm (accessed 01/14/13)

Wanda and I knew that the war would now affect us in some way or another. We didn't yet know the extent of our involvement but we did know that we hoped we wouldn't be separated for too long a period. Thousands of other couples most likely had that same hope while awaiting their calls to duty.

The Call Comes

On Wednesday February 23, 1966, I received my second and final call for induction. While I no longer have the original letter I received from the local draft board, the wording of the letter was as follows;

Selective Service System

Order to Report for Induction

To Eddie G. Nickels

Box 215

Whitesburg, Ky. 41858

Greeting:

You are hereby ordered to report for induction into the armed forces of the U. S. and to report at the basement of the

U.S. Post Office, Whitesburg, Ky. on 23, March,1966 at 5 am for forwarding to an Armed Forces Induction Station.

Signed----------- (member or clerk of local board.)

My destiny was now sealed. I would not be going into the Air Force, nor would I have to wonder about my plans for the future. I felt some relief now that I knew what was going to happen in our lives. No longer would Wanda and I have to plan around the unknown factor of whether I would be drafted or not.

Although Wanda and I were expecting my call to the service, we had never been able to decide whether she would try to keep on renting our little house or whether she would have to move in with her or my family, at least temporarily. Both our families lived close to our house, with my family's house only a few feet from ours.

I gave notice to J Don Collins that Friday, March 11 would be my last day of work so that I could have a few days off before I left home on the 23rd. Wanda and I had the opportunity to take a trip to Taylor, Michigan with my Dad, Mom, my sisters Rita and Kathy, and my brother Jimmy to spend a few days with Wanda's brother and my sister, Charlie and Marlene Adams. The main reason for the trip was to pick up my brother Phil who had been staying with Marlene and Charlie for a few weeks.

All seven of us loaded into my Dad's '54' Chevrolet and left Tunnel Hill the next day for Michigan where we stayed until Wednesday the 16th of March when we returned home. We enjoyed our visit to Michigan very much as Marlene and Charlie made us all feel very welcome on our visit.

Wanda and I also visited her brothers, James and his wife Henrietta (Rose) Adams and Paul and his wife Evie Adams while we were there. James and Henrietta took us sightseeing in the Detroit area and took us to visit Windsor, Ontario, Canada via the Ambassador Bridge and we stayed all night with them that night. We also enjoyed seeing my, and Wanda's, future brother in law Randy Rose while we were visiting. (Randy is Henrietta's brother and would later marry Wanda's sister Hazel.)

When we started back home we had to squeeze eight people into the car instead of the seven we left Kentucky with, as Phil was   also   returning   home   with   us.

Figure 5 Front row- Phil, Kathy, Wanda, Jimmy. Back row - Marlene, Nell, Rita, Eddie, Ed- Michigan trip March,1966

We enjoyed the trip home, especially since we all realized it would probably be our last vacation together as a family.

As we returned home that Wednesday I began to dread having to say good bye to Wanda and all my family in just seven more days. Wanda and I had never been apart for even one night in the one year and seven months of our marriage. I was determined to make sure that we visited all my relatives before I left home and to try to ease Wanda's mind about our separation as much as possible.

The weekend after we returned home Wanda and I started going by Dad and Mom's house and talking about old times and discussing the changes my induction would bring to our family. I also wanted to get a perspective from Dad about what Army life was like and what it was like to be drafted into the U.S. Army. Since Dad had been drafted in February, 1943 and knew what it was like to go through basic training, I hoped he would be able to prepare me what was to come.

He had never talked a lot about his induction and training during my childhood but had related a few incidents of Army life occasionally. He related to the whole family how he was pretty much running the operation of his Dad's truck mine when he was drafted and was probably eligible for exemption for the draft if he had chosen to contest it. Instead, he decided to do his part by answering his call when it came.

His Dad, Austin Nickles,[1] wanted Dad to go to his local board in Hindman, Kentucky and apply for exemption but Dad wouldn't hear of it. While my Dad was in basic training, Austin proceeded on his own to try to get Dad exempted as an industry critical worker but Dad refused the offer of a discharge. He always said that he was no better than anybody else's son that

---

[1] Dad's immediate family spelled their last name Nickles instead of the original spelling of the family name of Nickels that we used.

had to serve. He really felt that he wanted to do his part in the war.

He filled me in on how to get by in basic training and told some incidents of his own training. He said that when the war ended, his division had been training for months for the invasion of Japan and that very heavy casualties had been expected in his division by their Commanding General. He stated that everyone in his outfit was thrilled when the war ended and his division stopped their invasion training. He gave us his opinion as to why the Japanese had decided to surrender. He said," I guess they heard that Ed Nickels was coming to invade their country and they just didn't want to have to face me." That was typical humor from Dad.

My friend and childhood playmate, Thurman Sexton, came by one evening that weekend and related his experiences when he went through basic training at Fort Knox, Kentucky a few years earlier. He warned me about two hills in Fort Knox that the trainees hated with a passion because of the many times they had to tramp up and down them. He said they were known to the trainees as "Misery" and "Agony." That description put a little dread in my mind and I hoped that they might send me to basic training in some low land area instead.[1]

On Monday night March 21, Randall McNair came by Dad's to talk to me about the draft and Army life in general. He was a close neighbor that had just recently been released from the Army and he had a lot of helpful information for me.

He had actually been on a ship in The Pacific somewhere and on his way to Vietnam when he was pulled from the ship and

---

[1] Little did I know then that my hope and wish would come true.

sent home on emergency leave because of one of his parent's illness. This had happened when he had six months or less to serve of his military obligation.

Randal's perspective on military life was important, as were those of Dad and Thurman Sexton, because of his recent service. He was able to give me information on the current weapons and procedures that the Army was then using which would be important for me to know and understand.

By Tuesday, March 22, I felt I was as ready as anyone could be to face the unknown of military life. The information from Dad, Randal McNair, and Thurman Sexton had prepared me somewhat for what was to come. I was ready to go.

That is, I was ready to go except for dreading to leave Wanda. We had finally decided for Wanda to continue housekeeping, at least for now. We would take it one day at a time.

That evening we had difficulty even thinking about our parting and didn't discuss it much at all. I mentioned to Wanda that my previously high blood pressure problem might cause me to be sent back home from the Induction Center tomorrow night. We went to bed with that thought on our minds.

Chapter Five

INDUCTION

Answering Uncle Sam's Call

Figure 6 Eddie standing beside 1951 Mercury. House in rear was the one Wanda and I were renting at the time of my induction.

The morning of March 23, 1966 came much faster than I would have wished. We had gone to bed at eleven pm and I doubt that I had slept more than an hour or two at most that night. There was just too much anticipation of facing the unknown to have any semblance of a peaceful and restful night of sleep.

I thought it would make things easier if I could get out of bed and get ready to leave without waking Wanda. I slipped out of bed and got up to a very cold house and put my feet on the ice cold floor of the bedroom. Our fire in the coal heating stove located in the living room had gone out sometime during the night and I may as well have been outside as inside that cold morning. Since moving out on our own Wanda had got in the habit of taking care of the fire building chore and I probably would have smoked the house up pretty good had I attempted to rekindle the fire.

I got up at four am and sat down at the cold kitchen table and ate my usual two pieces of toast for breakfast. The electric cook stove (and a small radio to listen to the news) was one of the few amenities we could afford on my salary. As I sat there eating my toast the thought ran through my mind that maybe someday we might be fortunate enough to have some truly modern conveniences. I hoped that one of those conveniences might include some kind of better heat for our home.

It didn't take me long to pack a few things for my trip. A tooth brush, some toothpaste, a towel and washcloth and a change of underclothing was all I packed to take with me that morning. Every veteran of the military I had talked with had emphasized that no other clothing should be taken to the induction station. I had to defer to their knowledge on the subject as no instructions as to what to take with me had been included with the notice of induction.

Despite my trying to not disturb Wanda she woke up while I was eating breakfast and sat at the table with me as I ate. We talked for a few minutes while I finished eating and we both silently ached to the core of our being about having to part. It was a sad and defining moment in our young married lives as we knew when I stepped out the door our lives would possibly change forever.

We tried to act as normally as possible as I finished eating and picked up my paper bag with my few articles packed inside. It was as if we were trying hard to hold up in front of each other. As I walked towards the back door I struggled with my emotions and I knew Wanda was doing the same. I managed to put on a brave face and said to her, " Don't worry, If my blood pressure acts up at the examination today I'll probably be back sometime tonight!" Those words did little to placate her and as we embraced for the last time until no one knew when, I quickly turned towards the door, opened it and went out without another word and without looking back. It hurt too much to linger, so I hurried to Dad's car, started it and backed out of the driveway and drove down the dirt road to the main highway.

In later years Wanda said she watched as I drove down that dirt road and felt very alone and very concerned about the future and what it might bring. As it happened, she and I had the same thoughts about the unknown as I left home. Hundreds of other couples all over the U.S. were undergoing the same thoughts and concerns as they also answered their draft calls that morning.

I drove Dad's car to Whitesburg and parked it in the same parking spot I had used when I borrowed his car to have my first examination. I again slipped the key under the floor mat where I hid it the last time. If I failed the exam I would pick it up that night, or if I was accepted Dad would pick the car up later.

Figure 7   I parked Dad's car near the first telephone pole on March 23, 1966 as I left home for the Induction Station (Author's photo.)

As I walked towards the post office from my riverside parking spot which was located between the Salyers building and the old Elinda Ann Drive Inn Theatre, I wondered if I would be coming back home that evening or if I would be on my way to training at a military base somewhere. In just a few hours I would know the answer to my questions.

I arrived at the post office that morning at 4:50am and the draft board office was already open and checking in the draftees as they arrived there. As I looked around at the young men gathered there that morning I recognized several of them but knew only one of them by name. Jackie Sexton and I had been in grade school together at Whitesburg and were good friends. I was glad to see that at least I knew one friend that I would be in the Army with during our training.

There were a total of twelve men that checked in for induction on that Wednesday morning. That was the quota for our county that month, evidently. After our group checked in we were directed to go across the street while waiting for our bus that would take us to Ashland, Kentucky. A local veterans' organization had arranged for us to have coffee, juice, and donuts at the restaurant that was located on the bottom floor of the old Daniel Boone Hotel building.

I've always appreciated the fact that people cared enough about us that morning to get up before sunrise to see us off to the military. I had thought that we would just check in, climb on the bus and away we would go. I didn't think that anyone would even notice our leaving. They did this out of the goodness of their hearts, knowing that probably no one but us draftees would even know about their kindness to us. I wish I could remember every name of every man and woman that was there for us but I can only remember one person by name. Mr. August Codospoti, a World War One veteran, was front and center that morning. He thanked us over and over for being willing to serve our country. He and all the others who were present wished us all the best in our service to the country.

The bus pulled up about 5:30 AM and we were told to board the bus for our trip to the Induction Station in Ashland, Kentucky. As we went out the door we were offered a New Testament to take with us if we wished. I accepted one and was glad many times over the next few months and years that I did so, especially during the difficult times at Parris Island and beyond. Of course all this was in the future and it probably was a good thing I didn't know what was to lie ahead of me. If I had known, I might have been tempted to take "French leave" then and there.

Figure 8 The old Daniel Boone Hotel as it is today. (Tall building in center.)

As we once again traveled the same road to Ashland, Kentucky we had traveled when I went for my first physical in February, we had to endure smothering diesel fumes that made us all sick. The twelve of us spent the time discussing the training we would have to undergo in basic training at Fort Knox where almost all inductees from Letcher County were sent for their training. We didn't yet know that there were two of us on that bus for which that discussion was useless.

Just before noon we arrived at the large hotel building in Ashland, Kentucky where the Induction Processing Station was located. After reporting in we were lined up and again had to undergo a limited cursory examination to determine if our health was still as it was during our first examination. This exam wasn't near as extensive as the one we had during the pre-induction physical in February.

There were no written tests this time around and very few papers to sign. The only physical exam was a blood pressure check and a quick check of our hearts by listening through a stethoscope. That was about it except for one question that the examining doctor asked everyone. "Is your health still the same?" was the standard question. As far as I know everyone answered "yes" to that question. I expected to maybe have a problem with my blood pressure as before, but the doctor checked it and motioned for me to move on as he said to the next man in line, "Next!" I knew at that very moment that I wouldn't be returning home that night.

After our quick examinations we were directed downstairs to the dining room for a quick lunch before final processing for shipping out to receive our basic training. We still had no sure knowledge where we would be sent but we assumed it would be Fort Knox, Kentucky. At lunch we again discussed our limited knowledge of what the training would be like.

After lunch we were taken to a large room where dozens of metal chairs were set up in front of a raised platform with a podium in front. The twelve of us were seated together and we watched as a Marine Captain and two Marine Sergeants walked into the room in their sharp looking "Undress Blues"[1] and stepped onto the platform. They began looking through the folders of our Selective Service information as we watched them closely.

An Army Major then joined them and they stood there conferring among themselves for a short time. The Major then turned to us inductees and made a short speech about what an honor it was to serve our country, etc. After the speech he asked us to stand up, raise our right arm, and take the Oath of Enlistment into the United States military. The oath we took was as follows;

"I, Eddie Nickels, do solemnly swear (or affirm) that I will support and defend the Constitution of the United States against all enemies, foreign and domestic; that I will bear true faith and allegiance to the same; and that I will obey the orders of the President of the United States and the orders of the officers appointed over me, according to regulations and the Uniform Code of Military Justice. So help me God." [2]

The Major then informed us that from this moment forward we were under the jurisdiction of the United States military and we would be held to the oath we had just taken. With that admonition we were now Uncle Sam's boys .

After the swearing in ceremony, the Major looked out over our group of twelve new inductees and said, "We need two volunteers for the Marine Corps. Is there anyone interested in

---

[1] Undress Blues- Dress blue trousers with a red stripe and with a khaki or a "Tropical" shirt.
[2] http://www.history.army.mil/html/faq/oaths.html

volunteering for the Marine Corps?" Our small Whitesburg contingent of men looked around sheepish at each other as each of us hoped someone else might become the sacrificial lamb to the government. "Don't everyone volunteer at once, the Major interjected with a smile. You will only have to serve two years of active duty with the Marines, which is the same as the Army requires of its draftees. The Selective Service requirements are the same for the Marines as for the Army."

After all the inquiring I had done about Army basic training I certainly didn't expect or want to attend basic where I knew very little about their training procedures. I knew what to expect at Fort Knox because of all the former soldiers whom I had talked to in the days and weeks I was waiting for my final draft notice. This new development was a potential change in plans I certainly hadn't counted on. The Major would be waiting a while if he was waiting for my hand to go up, I thought.

Our group of men sat there stone faced and silent as the Marines and the Army Major looked us over and waited for a couple of hands to go up. I'm sure we didn't endear ourselves to the Marines on the platform when none of us raised our hands at that moment. Meanwhile I was thinking to myself that I certainly wouldn't be one of those picked. I was by far the most skinny and smallest in body size of our group, although I was one of the taller ones at just over six feet. I had been sure the largest and strongest of us would be picked to go to Parris Island.

As the group of military men gathered on the platform before us waited in vain for a hand or two to be raised, the Marine Captain walked to the Major's side at the podium. He reached into the stack of Selective Service folders lying before him and pulled out two folders. He then held them up as he said, "These two service folders belong to two of your group. When I call your names please step up here with us and follow these two Marine Sergeants to my office. We'll then get you squared away as to your orders and so forth."

My heart was beating with a vengeance as I sat there waiting for those two names to be called. I'm positive that every one of us sitting there were hoping that someone else's name might be called, but certainly not their own. The trip to (and through) Fort Knox was expected to be short and sweet. The trip to (and through) Parris Island was expected to be long and maybe brutal. Almost everyone I had talked to about Parris Island scared me with their comments about the training they had endured, or their friends had endured while attending boot camp on the Island. Nope, I wanted no part of that, thank you. Just send me to Fort Knox or home, if you please.

As the Captain held up the first folder he looked at our group and called out loudly, " Eddie G. Nickels, please step up here and go with the Sergeants!" When he called my name my heart was beating so loudly I was sure those sitting around me could hear it pumping. As I stepped up to the men on the dais there was a lump in my throat the size of a baseball. I didn't know if I should feel honored or royally screwed. After all, what were the chances that out of twelve men sitting there, my name would be the first one called? The choice of my name was purely at random it appeared. It could have been one of the meaner, tougher, and stronger of our group but it wasn't. It was me and my one hundred and thirty three underweight pounds.

I had always wanted to be in the Air Force, expected to be drafted into the Army, and now I had ended up in the Marine

Corps. I wouldn't have been too surprised if the Navy had grabbed me by the arm as I went out the door and whisked me to the Great Lakes Naval Training Center in Chicago. I didn't know what to expect anymore.

The Captain then called out another name," Travis Rickard please go with the Sergeants." This would be the last time for awhile that either I or Travis Rickard would be asked politely and with a "please" to do something for a member of the Marine Corps.

Rickard and I were escorted down the hallway to the Captain's office where one of the Gunnery Sergeants explained that we would be leaving the Selective Service Examining Station at 6 pm that same evening for the Huntington, West Virginia airport where we would catch a plane to Parris Island, South Carolina. (Actually our final flight destination would be Charleston, South Carolina.)

After explaining that we would have to wait on the van in an outside office and wouldn't be allowed to wander around until the van arrived to pick us up, the Gunny (E7) asked, "Now, are there any questions?" I spoke up and asked, " Sergeant I was kind of hoping to go into the Army with my friends. Do I have any choice in the matter, or am I required to go to Parris Island?" I didn't yet know it but an E7 Gunnery Sergeant is never addressed by the title of Sergeant but is called "Gunny". Also, a mere Marine private never addresses a Marine of any rank above private without a "Sir" being the first word of the address before that private has graduated boot camp. I hadn't even left the Captain's office yet before breaking several rules of Marine Corps etiquette and regulations.

The Gunny was patient and calm as he replied, "No, I'm afraid you have no choice in the matter. You two have been chosen or picked to represent your group in the United States Marine Corps. The decision has been made and is final. There's no choice involved here." It would be a long time before I was to hear kind and softly spoken words from a Gunnery Sergeant again.

After waiting in an office for a couple of hours the Gunny brought our orders to us which had finally been cut. He handed me our tickets and orders and said I was in charge of both sets of orders and tickets and would be held responsible for both sets of orders on the way to Parris Island.

Our orders included our Marine Corps Service Numbers which would be our identification the whole time we were in the Marine Corps. We would get paid by that number, we would be promoted by that number, and that number would be on our dog tags that would hang around our necks. We were instructed that we were required to memorize our own number by the time we reached Parris Island later that night. From that moment our social security numbers were meaningless to us as long as we served in the Marine Corps.

As the Gunny left the room we quickly opened the flap on the large manila envelopes and pulled out the page with the Service Numbers on it so we could begin memorizing them. When I looked at my number I could hardly believe my luck. My number would be a cinch to learn. As I stared at the page the numbers jumped out at me, 2233441. Except for the last number they were in sequence. Many times during boot camp I have listened as a "Boot" would sometimes struggle to remember his service number when asked by an inspecting officer. I never had that problem with such an easy number to remember.

At 5 pm we were taken to the dining area of the hotel for a quick sandwich while waiting on the van that would carry us to the airport. At 5:45 we left the dining room and walked to the street and climbed into the waiting van.

We crossed the Big Sandy River into West Virginia and proceeded to the airport on the mountain top in Huntington, West Virginia. While waiting for the plane to arrive I felt excited and somewhat concerned about the trip and what awaited us at the end of our destination. At the same time I just wanted to get the trip over with so I could face the next event in life. When we're young everything has an exciting aspect to it. As we grow older we tend to be satisfied to let others experience exciting things in life. This trip was exciting for me.

The Southern Airways two engine piston powered plane landed at 6:30 pm and we climbed aboard and into the two seats just in front of the wings of the plane. I had already claimed dubs on a window seat while Travis Rickard and I waited for the plane in the terminal building. I had never flown in a passenger plane before and I wanted to see everything and every building from the air that I could. Of course as it was dark when we took off, all I saw the whole flight was the lights of the countryside and cities but that also was exciting to me.

We soon took off into the night sky and as we climbed I was fascinated to observe the lights of the cities of Ashland, Huntington, Ironton, Catlettsburg, and other towns which were glowing and sparkling all over the area. It was a beautiful sight for one that had always dreamed of flying but had had little opportunity in the past to do so.

We landed in Roanoke, Virginia where a few passengers got off the plane and several more came aboard. We then flew to Charlotte, North Carolina where we changed planes for the trip to Charleston, South Carolina. This time we were flying in the first jet plane I had ever flown in. I was excited again! The thrill of the takeoff was as thrilling as a roller coaster ride to me.

After reaching Charleston, South Carolina, Rickard and I were directed to the bottom floor of the airport terminal where we were met by several Marine corporals and sergeants in their "Smokey Bear" hats. They were barking out orders left and right as several dozens of young men that were headed to Parris Island arrived from all parts of the Eastern United States.

We were given sandwiches to eat while we were waiting for the arrival of the buses that would take us on our final leg of our trip to Parris Island. We were also allowed to make a phone call home to inform our families that we had arrived in Charleston safely. The moment I called home was the first information my family had of my going to the Marines. When I told them where I was and where I was going they was incredulous and were very concerned about the change. Wanda wasn't at Dad's as she had gone home a little earlier after waiting until midnight without hearing from me. Dad said they would tell her that I had called.

After calling home I felt a little better about my situation. I had been wondering on the plane how I was going to inform members of my family that my supposed trip to Fort Knox had been changed to a different destination. Now I could relax a little since they knew where I was headed.

Since it was now after midnight the morning was now Thursday and the calendar had turned over to March 24, 1966.This new day would be the beginning of a new phase in life of which I had little inkling of at that moment in time. Most of us young men waiting on the buses that morning were unaware of the difficult conditions and training that awaited them on Parris Island. Ignorance was definitely bliss that morning.

The buses pulled in front of the terminal building sometime after 12:30 am and several dozen very tired and sleepy young men climbed the steps of the bus and slumped into a seat. There were a total of three buses and each of them was full as we pulled out of the terminal area.

I sit at a window seat near the back of the bus where I could watch out the darkened window as we left Charleston and proceeded down highway 17 towards Parris Island. Within a few minutes most of the passengers were napping quietly, talking in low whispers, or silently contemplating the events of the past day. Like Ritchie and me, most, if not all the passengers on the three buses were draftees who never expected to be headed for Parris Island for training. I'm sure many of their families were also surprised by their phone call informing them that their loved one wasn't heading for one of the numerous Army basic training centers for training.

Travis Rickard and I had somehow become separated as we boarded the bus and didn't sit together in the same row of seats. Instead I sat beside a young man from Boston by the name of Olson from Boston, Massachusetts. It seemed almost everyone else besides Rickard and I were from that city. We discussed briefly what might lie ahead of us when we reached Parris Island. We were both very apprehensive and concerned about what we might face in just a little while.

If I had known how long it would be before we would have an opportunity to catch some sleep I would have been trying very hard to catch a nap. However I was so wound up that sleep or napping was impossible for me that morning.

A little less than two hours after we left Charleston the bus driver spoke into the intercom and said," Gentlemen we're almost to the main gate of our destination". It was the first words he had spoken to us since we left the airport.

All conversation ceased immediately as we all looked towards the front of the bus. Ahead in the distance we could see a brightly lit area with a small building beside the roadway. We could see two Marines standing in front of the building and waiting for our buses to approach them.

Our bus slowed down and one of the smartly dressed Marines standing in front of the kiosk waved us on through. We were near the end our journey. The more difficult journey was about to begin.

I had left home that morning with the expectation that I would be going to Fort Knox which was fairly close to my mountain home. Instead, here I was, on a Trailway bus going to an Island training facility that I had heard so many disturbing stories about. I was hoping that those stories had been embellished by those who had been there before me.   I would find out soon enough that, to the contrary, those stories hadn't told half of what I would face over the next few weeks. The Marine Corps now had themselves a Boot from the Mountains, and   a   reluctant   Boot   at   that!

Chapter Six

RECEIVING

Our bus proceeded along the causeway and crossed a small bridge over Archer's Creek, then past a line of tall trees that ran along both sides of the road. We proceeded to go past the buildings and through the empty streets of the base. It was just before 2 am, March 24, 1966 and the base seemed eerily quiet at this hour of the morning.

The bus pulled in front of a large white two story building that seemed to be of World War One vintage. This was the "Receiving Barracks" where all recruits go through preliminary processing before their training begins.

We saw several Marines in the famous Campaign hats standing outside the door of our bus and the tension that we felt inside our gut intensified. The door opened and a Marine Sergeant stepped aboard the bus. He looked just as tough as one would expect a Marine NCO to look. I swallowed hard as he hurried aboard and glared at the young men as they anxiously and somewhat fearfully awaited his words." Now listen up!" he said in a loud voice." I welcome you to Parris Island on behalf of Major-General Masters Sr., Commanding General of the Parris

Island Recruit Depot. When I tell you to get off the bus I want you moving your ugly mugs off this fricken bus! When you hit the pavement you had best get your number tens into a pair of yellow footprints! Do you understand me?"

We managed to meekly answer, "Yes sir." The Sergeant then cupped his hand to his ear as he repeated, "Do you understand me?" Again we attempted to shout, "Yes sir!" "I'm asking you one last time!" he shouted. "Do you babies understand me?" When we answered him this time we almost made the bus shake as we screamed, "Yes sir!" " Then get off the frickin' bus now!"he shouted.

As we scrambled off the bus we hurried into a set of yellow footprints painted on the sidewalk. The Sergeant shouted for us to get at attention and we all did our best to stand as straight at attention as possible. It wasn't good enough for the three men in their Campaign hats that surrounded us. They proceeded to go through the rows of men as we stood in the footprints and screamed for us to "Stand straight!" and "Look straight ahead, don't eyeball me!" They would move from man to man as they chewed him out for the slightest movement of the eyes or body. A few shoves and "love taps" thrown in here and there accentuated their demands.

The same Sergeant that greeted us on the bus stepped to the front of our group as we stood in the footprints that formed our feet into a 45 degree angle. This was our first lesson on how to set our feet when at attention. Every movement we performed here after was planned to help us quickly learn Marine Corps procedures even when simply standing in line. Not a moment or movement was wasted. Everything was done for a reason and was geared toward teaching us the fundamentals of becoming a well disciplined and motivated Marine.

As the Sergeant lectured us he mentioned several articles from the Uniform Code of Military Justice that would guide our conduct while a member of the military. He made sure he emphasized Article 93 that prohibits disrespect to a senior officer and Article 91 that prohibits disobedience to a lawful order. He then asked us if we understood what he had told us and we again shouted, "Yes sir!" He stared a hole through us as he loudly repeated;" I said do you understand me?" We shouted as loud as we could this time, and it seemed to placate him for the moment.

The Sergeant then lined up our bus load of tired men and double timed us to a set of steel stairs (ladders in Marine jargon) that led to the second floor of the Receiving Barracks. We were led into a large room with rows of wooden chairs or desks with writing materials on them. We were told to pick up the felt marker and write on the back of our hand the number 381. Then we were told to pick up the ink pen and write on a red tag we had been given our service number, last name, and the platoon number that we had written on our hand. An hour or more was spent there doing paperwork.

We were then led a few at a time to a steel table where we had to empty our pockets of every item we had in our possession, including wallets, money, toilet articles, watches, necklaces, rings, and any other items we had on us. I had very little to empty out except for my billfold and my Bible I had received the day before when at the Daniel Boone Hotel in Whitesburg. I didn't have a single dollar or coin to turn in. My

billfold would be returned to me after training, I was told. My
Bible was thankfully and gratefully handed back to me.

I was shocked to see that two or three of the new recruits
with me had brought condoms and some pornographic
magazines with them. These were raked by a swing of an arm
into the trashcan by one of the NCOs in charge. Evidently some
of our group must have thought they were going on vacation or
something.

We were next directed to a room where we would receive an
issue of toilet articles, such as toothbrush, toothpaste, shaving
articles, brushes, deodorant, writing materials, socks,

towels, boot and shoe polish, razors, razor blades, and rifle
cleaning materials. (The razor and blades were items I could
almost have done without at that tender age.)

We were issued white cotton laundry bags in which to carry
our initial issue of PX items we had received. We were also
issued our "782 gear" consisting of a poncho, web belt, field
jacket, mess kit, canteen, and some other items. We did not
receive our clothing issue that first night.

After the PX issue we were taken to the top floor of the
Receiving Barracks where a long row of racks (bunks) were set
up. We weren't allowed to sit on the racks but had to stand at
attention. We were then told to "make a head call." Since I
didn't know what the term meant I stayed at attention. Most of
the men moved towards where ever the "head" was. When
they returned I whispered to the nearest man, "Where did you
all go just then?" He whispered out of the side of his mouth,
"To the restroom." The next time a head call was mentioned I
knew where to go. No one had informed us that the" head" in

the Marine Corps had the same function as a civilian restroom. I thought (foolishly) that every service called them "latrines." I had to wait three more hours for the call to come again.

We were made to fall outside and stand at attention again in the cold morning air. The field jackets we had been issued didn't help a lot because of the cold wind blowing off the ocean. The two noncoms with us tried to get us to stand straight as they went through our ranks shouting and showing us how to get our feet at a 45 degree angle and how to set our thumbs alongside our trouser seams. (There are no pants in the Marine Corps, just trousers.) "Shoulders back chin in!" they were shouting.

The Sergeant shouted "Attention", and we started marching, or a reasonable facsimile thereof. We rambled to the mess hall where we got our first Marine Corps lesson in how to hold our mess tray while going through the serving line. We held our trays against our chest as we came to each station where a mess man was ladling a portion of French toast and lumpy gravy, (S.O.S.) into our trays. The correct procedure is to hold your tray against your chest like a shield as you go through the line and hold the tray straight out when you are being served a portion of food. Your elbows are tucked into your sides and are kept there. We were allowed twenty minutes to eat our meal, no more and possibly a lot less.

No one could sit down until a table was surrounded by a full complement of recruits standing at attention with their elbows tucked against their sides. One of the recruits would then shout, "Ready seats," and everyone sat down in unison or you repeated it until it was performed in the correct manner.

Every bite you took had to be picked up with your fork or spoon and delivered to your mouth at a 90 degree angle. No cutting military corners here. When the Drill Instructor was finished eating you rushed for the door whether you were finished eating or not because you are required to be at attention and waiting in formation when the Drill Instructor stepped outside. We were given a little slack with our first meal because we weren't considered trainees yet.

After chow that morning we were taken back to the Receiving Barracks where I finally received a head call. We were ordered to "dry shave" with the razors we had been issued a few hours before. Some poor souls had reported for induction with beards and it was pitiful to hear them groan and holler. I think I could have dry shaved the whole time I was there without any problem.

Next, we received our haircuts which left us all bald. I couldn't even recognize Travis Rickard in the group of men after we had been shorn of our locks. We all felt like we had lost our identity with civilian life when we lost our hair. Of course that was the whole idea for shearing us.

After clipping us bald headed we were given our first class in hygiene and then led to the showers where the Instructors went along with us. We were instructed when to soap up, when to rinse off and when to put deodorant under our arms. Nothing is left to chance. It wouldn't do to have a bunch of smelly recruits running around on the base!

After showering we wrapped towels around our waists and were directed into a nearby room where our uniform issue would be made. We were measured and received four sets of "utilities" uniforms, plus uniform covers, (hats). Six pairs of "skivvies" (boxer shorts) and two sets of combat boots completed the clothing issue. All of this gear plus the PX issue we had already received were put into the seabag (duffel bag) we had been issued.

We then put our civilian clothes and shoes in a brown paper bag and rolled the top up and taped them for shipping home. We were told to throw our socks and underwear in the garbage.

We were then taken to the Administration Center where we filled out forms for our Servicemen's Group Life Insurance. Next, we signed statements swearing that all the statements and information we had given to the Marine Corps was accurate and truthful.

We also did a series of verbal, mathematical, and mechanical aptitude tests that would determine our MOS (Military Occupational Specially) while in the military. I requested duty as a clerk, a truck driver, or a warehouseman. I didn't expect to get either of my choices.

The next item on our list that day was a series of lectures on what was expected of us as new recruits during boot camp. The Sergeant lectured us on the stress associated with boot camp and the dedication it would take to complete our training. The pride we would feel when we were successfully graduated as fully fledged United States Marines would be worth all the sweat and effort we would be expected to expend during boot camp, the Sergeant explained. He also emphasized that we weren't Marines yet but we were mere lowly "Boots".

After noon chow we were run through another series of medical examinations similar to the ones we had taken at the Induction Station in Ashland, Kentucky. We were passed along from station to station as we were examined, poked, and prodded. When the Navy Corpsman took my blood pressure he made me sit down and took my pressure twice more.

After the third time he looked me in the eye and said, "Have you had any blood pressure problems in the past?" I then told him about the high readings from my first visit to the Ashland Examining Station in February of that year. He listened closely, then said, "Sit here while I summon the Doctor to check you out. I think you'll be going home instead of to training." My heart leaped with the joy of the moment but my hopes would soon be dashed.

The Navy Doctor came into the room and took my blood pressure several times and checked my heart again. He then looked at me and said, "Young man, I think you'll be alright. You're just under stress right now. You'll be fine for training." Oh joy, I thought. My hopes of heading home were dashed with those words. He outranked the Corpsman, to my regret.

We had our eyes and hearing examined again and I had no problem with those procedures. Those recruits that needed glasses were issued black framed military glasses for training.

After our examinations had been completed we were marched to evening chow. The mess men on duty verbally harassed us about being their replacements for mess duty and salutations of "You'll be sorry", as we went through the chow line. It had happened during our first meal on the Island and would continue until we achieved the appearance of being there awhile. New Boots always stood out like a sore thumb because of their wild eyed look and their body language. The cold weather we were enduring in March wasn't exactly ideal for achieving a tan it appeared and our pale skins hadn't seen much sun lately.

After chow we were formed in front of the Receiving Barracks and were instructed on some rudiments of military bearing, hand saluting, marching, and other procedures. We learned that the first word out of the mouth when addressing a ranking enlisted man on Parris Island Is "Sir". If you're asked a question and the answer is "no", the proper procedure is to say "Sir, no sir" or "Sir, yes sir," if in the affirmative.

Our first day at Parris Island had been a long day and an eye opener but we weren't finished yet. We attended several more lectures on Marine Corps history, etiquette, and traditions, among other subjects. We had no time for regrets or reflections about home.

After classes that evening we began having a "field day" in the barracks. To have a field day in the Corps is to clean everything from top to bottom. No speck of dust was left untouched. I managed to latch onto a swab (mop) and was at least able to stand up part of the time while swabbing. The swab handle was also handy for giving me an object to keep me from falling if I fell asleep as it was now past midnight Friday, March 25, 1966.

We got no sleep that night. I don't know why they even had racks in the barracks there in Receiving since we never got to use them. They even issued us sheets and blankets and went through the process of showing us how to make up a rack so that a quarter would bounce on them. I still can't figure that one out.

After field day we again stood at attention and were lectured and instructed on various things. At 0530 we were marched to chow and at 0700 we were sent on work details. My assignment was in the base laundry folding utility uniforms.[1]

We arrived back at the Receiving Barracks at 1730 and were marched to chow. I was so tired that I must have looked like a zombie, judging by how all my fellow recruits looked at me. I had not seen Travis Rickard since we had gotten our haircuts on Thursday. If I had I wouldn't have recognized him anyway.

After chow time came more standing time at attention and more lectures. I had still not been allowed to write a letter home since my arrival. A form letter had been sent home to let the family know I had arrived safely but there had been no opportunity to write a personal note. My return address had been included on the form. My return address was Pvt. E.G. Nickels, Platoon 381, S Company, 3RD.Recruit Training battalion, Parris Island, South Carolina 29905.

The letter was from the Commanding General of the Recruit Depot, Major General James M. Masters, Sr. (and was signed by Colonel James Juett of the Recruit Training Regiment.)It was sent when I first arrived on Parris Island. I had a chance to read the letter before it was sent home and to include my Parris Island address, but I had no chance to write anything else.

After the evening activities we were taken up the ladder (stairs) to the Receiving Barracks where those nice warm (and as yet unused) racks were awaiting our tired and worn out bodies. I don't how we were even still on out feet as we had been mostly standing attention in one form or another since our

---

[1] Utility uniforms were the working and combat uniform of the Marine Corps.

arrival early Thursday morning, except for the usual harassment activities and at chow time.

We had a couple of new NCOs in charge of us and we were in hopes of actually being taken to the barracks to sleep, but alas, our hopes were in vain. We were directed to field day the barracks again because "some dust had been found on top of my (their)hatch." So here we were, just barely able to function because of our fatigue, and engaged in cleaning the barracks from top to bottom. We weren't allowed to speak to each other while working unless we enjoyed doing pushups and crunches at midnight.

We spent a couple of hours to find that pesky speck of dust and the Sergeant said that " I've never seen such a sorry bunch of sloppy maggots in a barracks! I pity the Drill Instructors that will have to shape up such a sorry herd of misfits. You're all hopeless! I didn't know they were drafting girls now! You'll never be Marines, you're not fit to carry the title!" We had had some motivational speeches since arriving here but I don't think this would likely have been considered one of them.

We were directed into a room with several tables set up as if we might be given another lecture, but instead we were called to attention and were told to hold that position but to not lock our knees. (Locking your knees at attention would sometimes cause anyone doing so to pass out.) With those words I knew we would probably be standing there for awhile.

It was after midnight, Saturday, March 26, 1966 when we were ordered to assume the position of attention. I hadn't had any sleep, not even a catnap since I left home Wednesday morning, March 23rd. I was totally exhausted and my senses, reflexes, and muscles no longer acted with any coordination to speak of. After an hour or so of this activity my whole body was so weak I began to shake as if I had a chill all over my body. I wasn't alone as I could see several recruits across from me and they looked as if they were going to fall at any moment. I knew I looked at least as miserable as they did.

I heard it before I saw it. I noticed the sound of a thump here and a sigh there as several of our group of one hundred or so recruits began slumping to the deck. It seemed as though dozens went down during the next hour and I was in danger of joining them. I knew a human body can only take so much fatigue and stress before it collapses. I wondered in my mind how much longer I could last before I passed out also.

About thirty minutes or so after the collapse of some of the recruits began I noticed that there were only a dozen or so of us left standing. Oddly I thought, the NCO's weren't saying anything or demanding anyone stand back up at attention. They were just letting them lie there! Then it dawned on me. My mind was so foggy and my body was hurting so badly that I didn't realize at first that many of those that had fallen to the deck were completely faking it.

After the realization of what was happening I stood a few minutes longer and then sank to my knees and "fainted" away. I was asleep as soon as I hit the deck. It seemed as though I had just closed my eyes when I heard shouting and screaming coming from somewhere. When my head cleared I looked around and saw the recruits surrounding me getting to their feet. I then remembered where I was and jumped to my feet and assumed the position of attention once again.

A few slower recruits were trying to get to their feet as the Sergeant screamed, "Get the hell up you fakers, you're screwing up my Marine Corps!" At least I think that" fakers" was the word he used. I can't be certain though.

I don't know how long I was lying on the deck that morning. It could have been just a few minutes or a couple of hours. I do know that it helped some. At least those in charge hadn't been completely heartless. I've always wondered if there were any of us that were left standing that morning. If there were any they were better men than I was that night.

At 0430 we were marched to chow. The couple of days I had been on the Island I had come to hate the food and the whole mess hall atmosphere. The noise of the shouting and screaming combined with the noise of the trays, pots, and pans going through the scullery tended to increase my homesickness.

The food in the mess hall near the Receiving Barracks can only be described as tolerable. Maybe I was the only one that found it to be so, but that's the way I saw it. I would usually only eat the fruit offered and drink the orange juice. Thankfully the sign hanging in all boot camp mess halls that warned us to "take all you want but eat all you take" wasn't being enforced during our stay at Receiving. That would change when our platoon was picked up by our Drill Instructors whom would take us through our training, as we shall see later.

With chow over we were marched back to Receiving and told to ready our gear and seabags for moving out. We didn't know for sure what that meant but I hoped that it meant we were going to our training platoon. This waiting to begin training was for the birds. I wanted to get the training over with and go home on leave. I figured I was due a leave after spending the last couple of days on the Island! My impetuous nature was coming to the forefront even here at Parris Island.

We were herded out of the Receiving Barracks and into the yellow footprints on the pavement. We now numbered one hundred and ten green recruits in four files or squads as we lined up that Saturday morning.

As we stepped on the painted footsteps we were being shoved into formation by the four immaculately dressed Marines that were waiting for us at attention as we filed outside the Barracks. Their black loafers were shined to a high gloss, their web duty belts were gleaming with highly shined brass, their campaign ribbons were fashioned on their chests, and their Campaign (Smoky Bear) covers sat on their heads with the brim barely above their eyes. They were the best "Squared away" Marines I had ever seen. I had noticed many squared away Marines before and after coming to Parris Island, but these four Marines made a lasting impression on all of the new Boots of my platoon that day.

We were finally being "picked up" by the Marines who would escort us through boot camp. My first impression of them was one of awe, admiration, and fear of the unknown. My fears were well placed. We were in the hands of professional Parris Island Drill Instructors now.

Chapter Seven

FORMING for TRAINING

Enforcing Discipline

Figure 1 Colonel James G. Juett, Commanding Officer, Recruit Training Regiment, March, 1966 (USMC Photo)

After forming our group of recruits into four squads on the pavement the Drill Instructors placed one of their number at each corner of our platoon formation. When they had our alignment   arranged to their satisfaction the Drill Instructor with the Staff Sergeant's stripes and the black web duty belt shouted, "By the left foot, forwaaard, march!" We began marching the best we could along the road going to we knew not where. As we marched along the Drill Instructor made it easy for us by shouting left, right, left, right, left as we stumbled along. We didn't know it then but there would come a time a few weeks later when we would be able to keep the cadence and timing by the inflection of the Drill Instructor's voice, no matter what words he used in his cadence. They were keeping it simple during this first outing with us.

After marching twenty five yards or so along the street with our heavy seabags strapped across our shoulders, the Staff Sergeant shouted loudly, "Double time, hooooh!" We had already learned what double time meant in Receiving so we clutched our seabag straps tightly and began running and trying without success to keep in step with the formation.

It didn't take too many yards before the weight of that fully loaded seabag became a possible impediment to my successfully keeping in step and being able to finish this run. My load was almost more than my skinny legs could hold up, much less hold down while the seabag was flying first one way then another. Other recruits were having the same problem, especially the shorter ones who were having trouble keeping the seabags from dragging along the pavement as they ran.

We were in absolute misery after five hundred feet or so of pounding along the street with those heavy loads. Soon the inevitable happened. First one overburdened recruit dropped out of line and sprawled on the street and another quickly followed. As they hit the pavement a Drill Instructor jumped to the side of the first drop out and grabbed him by the scruff of the neck, pulled him to his feet, and gave him a swift kick to help him along. Meanwhile the other recruit was getting the same going over by the other Drill Instructor. As the kick was delivered the Drill Instructor was shouting, "Get along there Maggot, there'll be no quitters in my Marine Corps!" I couldn't help but think when I heard those words that the Marine Corps sure had a lot of Sergeants that owned the Marine Corps, since I had heard those words from our Sergeants in Receiving many times over the last four days.

Other recruits were falling out almost faster than the Drill Instructors could get them "cuffed and collared" and back on the street and running again. I think the punishment they were receiving when falling out was more agonizing and punishing than the agony they were suffering from the running.

When I saw what the recruits that either stumbled and fell or chose to fall out were receiving from the Drill Instructors I made up my mind to run until I got to wherever they were running us to or die trying. I sure didn't want a dose of what those unfortunate ones were receiving. That was medicine of too tough a dose to take for me. Of course that was the object of the Drill Instructors' efforts. If an example is made of a few the many will give their best effort and even go beyond their extreme limit of endurance most of the time. That's the Marine Corps way. It's tough, but effective.

After about a mile of running with reckless abandon I was wheezing and trying to gulp in more air than a 747 jet airliner as

I stumbled along. My long broomstick legs were pounding along the pavement and feeling as though they were holding up one thousand pounds each. Every time I would think I couldn't possibly take another stride I would think about the slap, slam, and kick that awaited me if I fell out. That thought would stimulate me enough for a burst of energy to keep going for a while longer. I felt like a sprinter who sees the finish line in the distance only to see the line moved back another one hundred yards when he approaches it. I was dying but my legs were still kicking.

Whatever one thinks of the methods used by the four Drill Instructors that day their method was very effective. All one hundred and ten recruits finished the one and one half mile run with those heavy seabags, which is a remarkable achievement for a group of out of shape and highly fatigued young men. Of course we can't discount the "help" of the Drill Instructors in making sure we all desired and obtained that achievement.

Our goal was successfully achieved when we reached the barracks. We were assigned to in the 3rd Recruit Training Battalion. This would be our home while in training at Parris Island. Some of those one hundred and ten men that survived that first hard run were destined to not graduate with this same platoon. Some would be discharged for various reasons and some would fail a phase of their training and be "set back" to join a new incoming platoon to complete their training. Nobody wanted to be set back in training. It would just prolong the time it took to get home on leave.

Although our platoon would lose almost thirty recruits from our platoon while training we also picked up some recruits that had been dropped back from other platoons. All in all, our platoon would have nearly as many recruits to graduate as we started with the first day of training. But that was in the future.

We reached the sidewalk in front of our new home in a terrible shape. Everyone was coughing, gasping, or throwing up as they tried to catch their breaths. I hadn't ran that hard since childhood when my sister Marlene was chasing me to whip me for some prank I had played on her, or even since a girl I was sweet on tried to slam me in the back with a schoolbook for calling her "Darling." Thankfully neither one caught me.

While the Drill Instructors were getting our group into a reasonable semblance of order on the sidewalk I was thinking how anxious I was to get into the barracks so I could sit down and rest. My legs were as wobbly as a newborn colt's. If I had known what awaited us inside those brick walls I would have gladly run all the way back to Receiving again instead of proceeding up those ladders to our second floor squad bay.

We shouldered our seabags once again and ran up two flights of concrete ladders and rushed into the large squad bay of our barracks for the first time. Our four squads ran to the end of the squad bay in single file as the Drill Instructors were screaming to the top of their lungs, " Get in front of a f......g rack you f......g scumbags. We jumped in front of a rack and stood at attention as the screaming and gesturing Drill Instructors ran back and forth grabbing first one recruit then another as they pushed and shoved them into position in front of the double racks. (Just like a set of bunk beds at home except much higher and more difficult to access the top rack.)

There was no rhyme or reason to the unforgettable happenings inside the squad bay that day. The object for the harassment was strictly for discipline purposes. Beginning in World War Two the requirement for training increasing numbers of men for fighting the war caused the training cycles to be shortened so the recruits could be quickly rushed to fighting units. There was no time to turn civilians into soldiers or marines by the long and slow methods of peacetime. With so many men to train, discipline was essential to good order and the best way to achieve that discipline was by quickly teaching new recruits that it would be to their advantage to get with the program and get on with the training. These tough methods of training had carried over with the Marine Corps since World War Two and were in place (at least in my platoon) when I was there. Some retired Marines swear things had changed by the 1960's and some others (like myself) swear they were the same. I'll just tell my story like I experienced it.

On April 8,1956 the Ribbon Creek Incident occurred on Parris Island when Staff Sergeant Matthew McKeon led 74 recruits on a late night march into the swampy creek while subjecting the platoon to extra disciplinary measures. This exercise resulted in the drowning deaths of six recruits. After that incident boot camp was revised and restructured to help insure that such an incident would not occur again. The number of Drill Instructors (Drill Instructors are never referred to as DIs) per platoon was increased and other changes were put into place. This change was supposed to curb the discipline problems.[1] You wouldn't know any difference by my platoon's experience though.

---

[1] http://en.wikipedia.org/wiki/Ribbon_Creek_incident

It was now 0900 on Saturday morning and the run we had just completed from the yellow footprints of the Receiving Barracks to our squad bay in 3rd Battalion had worn us completely out. My knees were shaking with fear and fatigue far worse than they had when I had "fainted "and fell asleep just after midnight this morning.

The recruits were at attention along both sides of the squad bay as the Drill Instructors were going up and down the line picking out first one recruit, then another to harass and "impose discipline" on. The shouting, screaming, crying, cursing, swearing, groaning, and other noises were a deafening roar in the squad bay. I and the other recruits were feeling fear, confusion, shock, anger, and fatigue all rolled into one gigantic ball of craziness as we stood at attention and wondered what was happening and what would happen to us next.

I wasn't able to observe much of what was going on except for what was happening in front of my 'locked to the front' eyeballs or what was happening beside me. By the noise level and all the shouting, cursing, and screaming I knew unusual things were taking place. First one Drill Instructor would fly in front of me and put his nose on mine and shout, " Don't you f....ing eyeball me again screwball, do you understand me?" I would weakly shout "Yes sir!" That would bring a tirade from him, "What's the first f...ing word out of your mouth boy?" I would then repeat, "Sir, yes sir!" With those words he would move to the next one to plant his nose on their nose for some perceived offense. A few blows were also delivered liberally.

I would hear the screaming rise to an almost deafening crescendo at times and see bodies flying all over the place. The reason bodies were lying here and there wasn't known for sure but evidently some movement had been made outside the boundaries of the rules, such as moving your eyeballs or showing fear in your eyes.

No pen and paper nor the greatest orator in history could satisfactorily tell the events of that morning in a way that could be understood by those that have never gone through the rigors of Marine Corps Boot Camp. Those events, no matter how harsh they were to endure, will always bind together those that have experienced them. I have tried to remove most events from my mind over the years. Those incidents mentioned here gives an idea of our ordeal. You would really have had to experience it to understand that the price of freedom is bought at a huge price, even in training. None of our group was prepared for the mental and emotional stress we suffered that first day with our Drill Instructors. I doubt that anyone could really be prepared for that kind of stress. After that experience potentially being captured by an enemy during combat held little fear among my platoon.   This was our initiation into the fraternity that had bound together Marine Corps veterans since the very beginning of the Corps at Tun Tavern in Philadelphia, Pennsylvania on November 10, 1775.

We had been standing at attention two hours or longer when one of the Drill Instructors, a Corporal, jumped into my face and accused me of eyeballing him. If I did eyeball him as he passed by it was because of fear, but as I recollect I was making a mighty effort to not move a muscle or eyeball, so the issue was in doubt. Putting his nose on mine again he proceeded to shout to me his doubt of my being anything but the lowest slimeball that ever lived. Then, after he finished his tirade of my kinship with dogs he grabbed me by the neck with both hands as we both went flying to the bulkhead (wall) and hitting the side of the rack as we fell to the deck. I instinctively reacted by grabbing his beefy hands and trying to pry them from my scrawny neck, but to no avail. In a few seconds the room started spinning and the lights went out.(For me anyway.)

I have no idea how long I was lying on the deck but it was probable a couple of minutes at most. When I started to regain my senses and lifted my head from the deck to look around I saw the rows of racks and thought I was in a hospital setting. Why was I seeing the legs of so many people standing in front of the hospital beds, I thought? Then I realized I was hearing shouting and other noises in the background which cleared my head enough to remember how I had ended up on the deck. I quickly debated in my head whether I should just lay there or whether I had better stand at attention again. My sense of duty and fear of even harsher consequences prompted me to grasp the side of the rack near me and pull myself back on my feet.

My neck was still burning and hurting which made me realize how the old frying chicken must have felt when Granny was swinging it around and around by its neck to kill it for supper several years ago. I felt somewhat of a kinship with that old chicken at that moment. I could only hope that the Drill Instructor that throttled my neck hurt his hands more than my neck while he was squeezing it. I was just going to stand at attention in front of that rack again and pretend that nothing happened at all and I wasn't hurting a bit. With that thought I rushed to the front of the rack and assumed the position of attention once more.

85

All kinds of swearing, screaming, and shouts were still reverberating throughout the squad bay. The squad bay had a line of racks on each side of the deck and each double rack had two recruits standing at attention in front of it. As I stood there braced with my shoulders back and my chin tucked close to my body, I could see the fear in the eyes of the recruits at attention across from me. I knew I had the same look that they had or maybe even worse. The Drill Instructors were still engaged in going from recruit to recruit with their harassing tactics. How long was this act in the play going to last, I mused. We had been enduring this for at least two hours and each minute seemed like an hour as we waited for the next hammer to drop.

As I stood there I was surprised to recognize my friend from Letcher County, Travis Rickard, standing across from me. He was standing at attention about two racks up from mine. It wasn't long until one of the roving Drill Instructors stopped in front of his rack to pay him a visit.

I won't give all the details in his situation, (I was too scared to watch closely,) but I will say it wasn't long before I watched out of the corner of my eye as he hit the deck. I could see he wasn't moving and I thought that he might have been knocked out by the fall and would soon come around. I tried hard to look straight ahead but I could still see him lying there after a couple of minutes. I don't know if his passing out was due to stress or illness or some other incident. It was likely the latter.

Soon another Drill Instructor came by and hurried to his side and a few minutes later Rickard was helped out of my sight and out of the building. It's not clear to me if he was taken by a Drill Instructor or whether he left by ambulance. I heard later he was admitted to the hospital. The forty seven years that have passed since then have dimmed my memory of this incident. There was so much other activity going on around me that the whole scene was a blur of surrealism at the time.

Two days later Rickard was escorted back to the squad bay to pack his gear. I was standing at attention as I watched him but all he could do was look at me from across the aisle and mouth the words "going home." I envied him as those words sunk into my mind, but I was also glad for him. I could only hope that the cause of his release was nothing serious. I haven't seen nor heard of him since that day.

We were soon prompted to "Get the hell outside and get lined up for chow!" Get out, they shouted, run, run, move, you've got one minute to get in platoon formation!" We raced down the ladder as the Drill Instructors helped us along by screaming in our ears as we went out the hatch. (There are no doors in the Marine Corps, just hatches.) We got into a rough facsimile of platoon formation and we were then marched to chow. I use the word march loosely here. In a few days we would be able to march with the best of them.

Going to chow was an adventure. We lined up in platoon formation, four squads of just over twenty five men each, standing quietly at attention as we advanced into the line leading into the mess hall. The four men in front would step forward on the command by the Drill Instructor, "Step! Face! Uncover! Move!" With the four squads of our platoon standing perpendicular to the concrete walkway leading into the mess hall we would take one step forward on the command "Step", turn to the left on the command "Face," remove our covers (caps) on the command "Uncover," and start moving forward on the command of "Move."

As we entered the mess hall we each removed a tray from the tray rack and then picked up a knife, fork and spoon from the utensil containers. The aluminum tray is held closely to the chest as we begin side stepping along the serving line. At the same time the back is straight and the head is pulled into the chin as closely as possible, just as if we were assuming the position of attention. The elbows are tightly pressed against the sides of the body while holding the tray at a forty five degree angle. As you move from one serving station to another you are facing the servers. You move to the next server by sidestepping. The right foot is moved to the right and the left foot follows and clicks together with the right foot.

Once you have reached the end of the line and have your food you gather around the long tables and stand at attention with your now full tray being held at against the body with your arms and elbows tight against the body. When a full complement of hungry recruits surround the table an observing Drill Instructor shouts, "Ready, seats!" You are then allowed to sit down and begin eating.

Each bite you take had to be eaten in a military manner. The fork or spoon is held straight out and goes to the tray in a straight motion and the food on the fork is raised up in a ninety degree angle. Then the food is brought to the mouth in a straight line while the elbow is still tucked tightly to the body.

There are no round corners in a Marine Corps Boot Camp. There are straight angles only.

When you sit down your back has to be perfectly straight as you sit only on the first four inches of the bench or chair and with the feet at a forty five degree angle flat on the deck. There's no talking, looking around, or "gawking" as the Drill Instructors call it.

If a glass of milk is desired, one is required to stand at attention and shout, "Sir, this recruit requests to "milk the cow", sir!" If the request is granted the recruit proceeds to the line leading to the dispenser where the large milk cartons are located and lifts the lever to dispense the milk. He then hurries back to the table and reclaims his four inches of the seat and assumes his straight angles only position while he is eating.

As the recruit finishes the meal he takes his tray to the exit hatch where two shiny, large garbage cans sit side by side. One can is used for raking any uneaten food into it and the other is full of hot water that is used to give your tray a rinsing before taking it to the scullery to be washed and dried. The garbage can that the scraps are scraped into is always nearly empty as you are required to "Take all you want but eat all you take." A Drill Instructor is stationed at the garbage cans to enforce that rule. Any violators are sent back to a table where a Drill Instructor observes to assure that every bite is consumed.

A recruit is allowed twenty minutes for each meal and is required to be outside in platoon formation before the Drill Instructors finish eating. Woe beyond the whole platoon if the Drill Instructor has to wait on a recruit to return from the mess hall. While waiting outside the mess hall you must have your "little red books" open and reading and studying your eleven general orders, your chain of command, or other procedures or regulations.

After our first 3rd Battalion mess hall meal that Saturday, we were marched back to the barracks where we once again assumed the position of attention. After the events of that morning I was dreading the afternoon and night. I wasn't disappointed, as the verbal and physical altercations continued all day and into the night, but the real tough incidents got fewer and fewer as the day progressed.

The only other incident of note occurred when one of the recruits lost his cool and applied his fist to the face of one of the Drill Instructors while the recruit was suffering a berating.

The recruit was standing at attention about five feet from me and I heard the blow and saw the Drill Instructor hit the deck and slide across the deck. He then jumped from the deck as he pointed to the offending recruit and shouted, "He hit me, he hit me!" At that moment all the Drill Instructors rushed to the offender and hustled him to a port (window) where they grabbed him by the boots and legs and proceeded to hold him out the port. I think that shook him up as it would any normal human being and caused him to cry out. After a short time the recruit was pulled back inside and shortly two MPs appeared and escorted him out of the building.

He stayed gone two days before he was escorted back into the squad bay again and where he was those two days was only speculation only so I won't repeat it here. I later got a chance to ask him about the incident but he said he preferred to not talk about it. I do know that he not only graduated with our platoon but graduated with Honors.

Our ordeal continued until sometime into the night when we were instructed how to make up our racks correctly and each recruit was assigned a specific rack. I ended up being assigned to the same side of the squad bay where I had been at attention all day but my assigned rack was only ten feet from the desk where the Drill Instructors would sit while instructing and watching us. That would be my home until graduation and I was never able to skirt the rules very much because of my position near the Duty Drill Instructor. The Duty Drill Instructor never left our side or sight until lights out when he would enter his office for the night, which was only fifteen feet from my rack.

The two rows of racks were called the port side and starboard side racks. "Port " is left and "starboard" is right in naval parlance. I was assigned to the port side of the squad bay. We would learn to fall outside by the port side and go to the head by the port side. We went by the group, even to make a head call. We only had four or five minutes allowed for group head calls but somehow we managed to achieve our goals and were back at our racks in the allotted time.

That first night we were given a lesson in Marine terminology that was required to be used during boot camp and throughout our Marine careers. "It's a *deck* where you're standing, not a floor. We go *down below*, not down stairs. We don't go ladders, we go *topside*. The thing behind your racks is a *bulkhead*, not a wall. In those bulkheads are *ports* not windows. Behind me is the *head*, not the latrine or toilet," etc., etc.

After the members of our port side were led into the head for showers by a Drill Instructor we used our four minutes to scrub off the best we could and were directed to tie our towels around our waists and to stand at attention by our racks. After the starboard side had their turn they likewise stood at attention as we were instructed to open our seabags . Each item inside the seabags was pulled out and inventoried in front of the Drill Instructors to make sure we had every required item.

After the inventory each item was folded and put neatly back into the seabag or stowed in the foot locker which was stored under the rack. Spare utility uniforms were stored in our footlockers along with the extra pair of boots and running shoes.

After the footlockers were stowed under the racks we stood at attention as the Drill Instructor shouted, "Hit the racks!" With those words I planted my left foot on the bottom rack and vaulted into the top rack I had been assigned to. The tall recruits were generally assigned the top racks to facilitate getting into bed quickly and not allowing the rack to squeak as they did so because any noise when mounting the racks would cause many repetitions before the Drill Instructor was satisfied. That first night it took at least thirty minutes to please him.

I was totally fatigued and weak as water when we were finally allowed to hit the racks that night. The day's events had drained all of us of the last ounce of energy in our bodies. If every day was like our first day with our Drill Instructors I doubted my ability to hang on in the weeks to come.

The events I had witnessed and experienced that day were almost traumatizing to me and all of my fellow recruits. We had never dreamed it could have been as rough as it was. I knew I would never be able to fully verbalize to anyone what we had been through. I was not even sure I would ever even want to tell anyone about my experiences. Even today it's painful to pull up those memories.

The introduction to our Marine Corps Drill Instructors was now past us. It had been even rougher than I had imagined or had heard. I felt proud that I had weathered the first storm.

I hadn't been in bed since Tuesday night March, 22. It was now Saturday, March,26. I was asleep as soon as my head (and sore neck) hit the pillow.

It seemed as though I had been asleep for only a few seconds when a loud banging noise woke me up. I looked towards the loud noise just in time to see a large garbage can sailing into the middle of the deck of the squad bay. A Drill Instructor was violently banging two metal garbage can lids together. The four Drill Instructors were shouting and getting in our faces as they shouted, "Get dressed, hurry, hurry, you've got one minute! You're too slow girls! Drop down and give me twenty five pushups hurry, hurry!" One hundred pushups and fifteen minutes later we were still trying to get all our clothes on. We would get one leg in our trousers and they would shout "Freeze." Looking around, they would observe several recruits not completely dressed and would drop the whole platoon for more pushups. They would then allow one more minute to try to finish dressing and when "freeze" was shouted again and everyone wasn't completely dressed we'd drop for pushups again. It must have taken at least twenty or thirty minutes that morning to get fully dressed into our utility shirts and trousers.

We were then marched to chow and once again endured all the rituals it took to get through a meal at Parris Island. I was feeling very ill and home sick that morning and was glad to see that a light breakfast was on the menu.

```
                THE BATTALION CHAPEL
        THE THIRD RECRUIT TRAINING BATTALION
            PARRIS ISLAND,  SOUTH CAROLINA

Lt. Col. G. L. Lillich.  .  .Commanding Officer
Major F. J. Werz.  .  .  .  . Executive Officer
Lt. R. J. Paciocco . . . . . . . Chaplain
Mr. J. Caldwell . . . . . . . . . Organist
***********************************************
0900            27 March 1966            1030
***********************************************

                ORDER OF WORSHIP

Prelude                              Organist
Call To Worship and Invocation       Chaplain
*Hymn   "When Morning Gilds The Skies"  No. 115
*Hymn        "The Old Rugged Cross"   No. 210
Presentation Of Our Tithes
*Doxology                            No. 418
The Morning Prayer                   Chaplain
Sermon                               Chaplain
*Hymn       "Soldiers Of Christ, Arise"  No. 300
*Benediction                         Chaplain
Postlude                             Organist
        A Service of Holy Communion for
        the graduating Series will follow
            our 1030 Worship Service.
***********************************************
        *The Congregation will stand.
***********************************************
Tear off & place in offering plate if you wish.

____I wish to speak with the Chaplain.
____I would like to take Rel. Instructions.

NAME:_____    PLT._____
```

Figure 2 March 27,1966 Church Bulletin- Parris Island   (Authors Collection)

Chapter Eight

PHASE ONE CHALLENGES

Physical and Psychological Training

I am the last recruit in the first row looking over the Drill Instructor's shoulder. USMC Photo- March, 1966. The Drill Instructor is Sgt. Shue, the "Heavy Hat" or second in command of Platoon 381,3rd Battalion, S Company, Recruit Training Regiment. He was responsible for teaching much of the academic knowledge and also responsible for the overall discipline of our platoon.

At 0330 hours on Monday March 28, 1966, our platoon again awakened to the sound of garbage can lids being slammed together and garbage cans hitting the deck while the drill Instructors were shouting commands to the top of their lungs. Although we were expecting that loud wakeup call of Sunday morning to be repeated it was still shocking to experience it.

We were once again given one minute to dress and predictably we had to do pushups and side straddle hops until everyone got dressed. Afterwards, each side of the squad bay made head calls, first the port side, then the starboard side. Going to the head in groups of over fifty men would take some time to get used to but after two weeks of the routine it would seem like we were going to the bathroom at home. Of course at home I was used to an outside toilet so I considered myself lucky to have the head so close to my rack every morning.

After chow we proceeded to the 3rd Battalion "grinder" or parade deck where we were destined to have close order drill many, many times while at Parris Island. We spent some time with the Drill Instructors assigning each of us to a squad in the platoon. The members of each squad were picked to occupy their permanent positions by height. The tallest men were put in the front of each squad and the next tallest would fall in behind him and so on all the way down the line.

Four squads were lined up in their own file, with approximately twenty five men in each file or squad. My assignment was the third squad, the fourth man in the file. That meant I was the fourth tallest man in my squad and the fourteenth tallest of the one hundred and ten man platoon. This number would change by the end of this day as some of our number would fail their Initial Strength Test. (IST)

Our platoon was joined on the grinder for the first time by the other two platoons in our series, Platoons number 380, and 382. My platoon, 381, was in the middle squad bay of the barracks while 380 was on the first deck and 382 was assigned the third deck of our barracks. All three platoons made up our company, which was S Company of the 3rd battalion. We would be training and graduating together, providing we managed to get through boot camp.

Phase one of our training would last approximately three weeks. This part of our training was designed to build up our bodies and to break each of us down psychologically. That part had started on Saturday in the barracks.

We would quickly learn not to use the first person when talking to a drill Instructor. Me, my, I, and you was forbidden to use in boot camp. When addressing another person of any rank we had to refer to ourselves as "this Recruit." For example, when addressing a Drill Instructor I would say, "Sir, this Recruit requests permission to speak to the Drill Instructor Sir!" The Drill Instructor would then say something like, "What do you want dummy?" Another favorite phrase was "Speak Turd!" Sometimes it was, "Get out of my face, Shitbird!" If the last phrase is used you probably won't receive permission to speak.

To use the pronoun "you" was the worst sin you could possibly commit in front of a Drill Instructor. The first week at boot camp I had an occasion to find that out the hard way. I approached the Drill Instructor at night in the squad bay and asked in the loudest voice I possibly could," Sir, this recruit requests permission to speak to the Drill Instructor, Sir!" He looked at me while scowling and shouted, "Speak, shithead , and it better be important!" I swallowed hard as I again shouted, "Sir, this recruit would like to ask you if this Recruit can make a head ca…." That's as far as I got. He grabbed me by the buttoned up neck of my utilities as he screamed, "A *you* is a female sheep numb nuts! Do I look like a female sheep, Chrome Dome? I gulped and swallowed a couple of times as I screamed a reply, "No Sir, the Drill Instructor definitely does not look like a female sheep, Sir!"

With those words he released his grip on my shirt collar and shouted, "Now is this head call an emergency?" I replied, "This recruit doesn't know for sure Sir, but it might be pretty soon!" He glared at me as he shouted, "Well then dipshit, get the siren going and get a move on!" I then took one step backwards, clicked my heels together, then screamed as loudly as I could, "Aye aye, Sir," did an about face and took off running. I circled the entire barracks while making a screaming noise like a siren with my arms held out like airplane wings. I was shouting "Whinnnnnneeeee" as I circled the inside of the barracks on my way to the head. As I passed the Drill Instructor at his desk I had the siren going as loud as I could so I wouldn't be flagged down by him. I don't think I ever made the mistake of using the word "you" again while in boot camp.

The reason for the aversion to using the first person in Boot Camp is to break the recruit from his individualistic habits he or she has acquired in civilian life and to think of themselves as part of a team. Teamwork is vital to accomplishing almost impossible tasks in wartime and this teamwork begins in boot camp.

We spent all morning of the first training day learning how to march and drill as a platoon. By noon we were getting to become familiar with the different commands and how to execute them. We didn't have our rifles yet but we had enough trouble learning how to march that first day, much less having to learn the manual of arms at the same time. Drill, drill, drill, cadence, cadence, cadence, it was all new to us but in time it would feel and seem as natural as walking, to our platoon.

It would be hard to imagine a shabbier, ragged, and scared platoon of misfits as we must have looked that first day of drill. Almost every one of us found ourselves out of step more than we were in step and "diddie bopping" along was a commonality among our squads. Our utility jackets were buttoned all the way up to our adam's apples so tightly we could hardly breath and our trouser cuffs were unbloused and hanging loosely over our boots. Buttoning the top button and leaving the cuffs of the trousers unbloused was required of all new recruits. Those two little tidbits were like everything else at Parris Island. They had to be earned. Only when you had sufficient time on the Island, when you were sufficiently competent in drill, and when you have shown a deserving attitude, were you allowed to look anything like a Marine. After all we weren't anywhere near worthy Marines yet and we wouldn't be until the day we received that Eagle, Globe, and Anchor emblem and graduated recruit training. Until then we were simply Boots that deserved nothing, not even the respect of those who wore the proud title of United States Marines.

Around noon we were marched to the nearby mess hall for chow. Then we marched back to the barracks and were ordered to take off our utility blouses (shirts) and change to sweat shirts. In March it gets mighty cold on Parris Island as the wind blows in from the ocean and the sweatshirts were some warmer than the utility jackets and they made running more comfortable.

We then marched to a large grassy field where we would take the Initial Strength Test. This exercise was designed to weed out those that were too weak or too slow to start the training cycle. Those that failed would be immediately transferred to Physical Conditioning Platoon (PCP).They would no longer train with our platoon and would be merged with a newly arrived platoon after their physical condition was up to par.

I was plenty worried about taking the test. I certainly didn't want to think about having several extra weeks of physical conditioning before I started training. That would only prolong my stay on Parris Island. I was so homesick that I just wanted to get on with the training and get home on leave. (I had been gone from home for only six days at this time.)

All three platoons of "S" Company would take the test which consisted of a minimum of three pull-ups, thirty five sit ups in two minutes, and a one and one half mile run in 13:30. It took most of the afternoon to take the test as there were more than three hundred recruits in the company.

When my turn came to take the test, my first event was the pull up bar. We were required to hang from the bar by our arms and pull ourselves up until our heads were above the bar. This required arm strength alone and I knew this would be my weakest event. I managed to do two pull ups but on the third one I really struggled to get my head above that bar. The third one was so debatable as to whether I had done it correctly that the physical instructor made me do another one. I jumped up and grabbed the eight foot high bar and strained like the dickens to get my head above that doggone bar. After what seemed like an eternity I finally got my complete head and neck above the bar. I imagine that the veins and arteries in my neck looked like ship mooring ropes as I strained to lift my head above that piece of cold steel.

My next test was to do the thirty five sit ups. That was my favorite, as I managed to do fifty four in the two minutes allowed for that event. My skinny body came in handy on that exercise.

The one and a half mile run was tough on me even though I was accustomed to walking nearly everywhere I went at home. I found that walking and running were two different things. I came in with a time of just over twelve minutes, but still I had struggled to make it.

I was very pleased to have passed the first physical hurdle of boot camp but I realized there would be many more hurdles to come. Several recruits failed the tests and were transferred to P.C.P. that evening after chow. There were three or four in my platoon and about that many from each of the other platoons. After our first test our platoon was already losing recruits.

The rest of the day was spent in physical exercise and drill. After evening chow we continued drilling and many recruits were called out for extra punishment because of some kind of infraction. I was fortunate to be sandwiched into the third squad where I could get by with a few mistakes in drill but I still did my share of "incentive" exercises nearly every day.

At the edge of dark we marched inside the squad bay to continue our training by listening to lectures and procedures while standing in front of our racks at attention. There was no such thing as sitting down in the evening until around 2100 hours when our so called "free time" period began. That was when you could pull your foot locker out from under your rack to sit on while you read manuals, studied your "Guidebook for Marines," cleaned your rifle, and wrote letters home. This first night we were too busy to write letters and had not yet been issued rifles but we had plenty of brass to shine and boots to polish.

At 2200 hours "lights out" was ordered and once again it was thirty minutes or so before we had satisfied the Drill Instructors enough to not have to jump back off the rack and try to mount it again without a mistake or a noise of any kind. I again was asleep instantly.

We had come into the barracks that evening to find the contents of our foot lockers dumped on the deck and foot powders covering every item we had in the lockers. The sheets and blankets were torn off the racks and the mattresses were scattered on the deck of the squad bay. It looked like a bomb had made a direct hit on our belongings. We quickly cleaned up the mess and got it in an order that (almost) pleased the Drill Instructors in an hour or so. Of course they were very helpful by the use of their verbal 'encouragement' as we labored. There's nothing better suited to encourage perfect compliance than a Marine Corps Drill Instructor's shouting into your ear while his mouth is only one inch from your ear. It works every time.

The usual garbage can noise woke us up again at 0330 hours the next morning. After our four minute head call and the usual harassment while attempting to get our clothes and gear on we proceeded outside and into formation. Some appropriate physical exercises followed, then we were marched to a large field where we started drilling to the cadence of the Drill Instructor. After a few steps we heard the words " Double time Hooohrrr!" We then ran our first of many three mile runs in our cumbersome combat boots. Several fell out and had to be helped by other recruits to complete the run. We had already been warned that nobody was to ever be left behind on a run or an exercise of any kind. We would work as a team no matter what the job was. If a man falls out the closest man carries his equipment or helps him complete the run or march. No excuses were allowed except for heat stroke or unconsciousness.

We were fortunate that morning that we hadn't yet been issued rifles and field packs. We were completely gassed at the end of the run. Not many of us were able to enjoy chow after marching back to the 3rd Battalion area and to the chow hall. At 1700 hours that evening we would again have a three mile training run. This time we would have rifles to carry at port arms as we ran. We ran that three mile run every day at least twice while in training except for Sundays and light training days when the temperature reached 90 degrees or more. On the 90 degree days the "black flag" was raised which restricted the amount of physical activity allowed. I saw very few of those black flag days during my time in training.

After chow we were marched to a warehouse where we received our "bucket issue" which consisted of an eight quart size galvanized pail, some scrub brushes, a shelter half, some tent poles, haversack, knapsack, canteen, canteen cover, canteen cup, and tent pegs. We would later spend many hours learning how to assemble a light marching pack and a field transport pack.

After the bucket issue we were marched to the armory where our soon to be familiar M14 rifles were issued to us. As we lined up and filed through the armory we saw several large gun racks loaded to capacity with M14's. I had grown up with a hunting rifle in my hands and was anxious to learn all about this fine looking weapon.

The M14 was adopted by the U.S. military in the early 1960's to replace the M1 Garand of World War Two and Korea fame. While the M1 was a fantastic weapon in its own right, the M14 had several improvements over it. The M14 had a twenty round magazine while the M1 had an eight round clip. The rate of fire for an M1 was 16-24 rounds per minute. The M14 rate of fire was 700-750 rounds per minute when the selector on the side was switched to "FULL." (The M14's in boot camp had the selectors removed and could fire only in a semi- automatic mode.) The M1 fires a .30-06 cartridge while the M14 fires a 7.62 NATO cartridge which is slightly smaller and lighter than the .30-06 round. This allowed more 7.62 cartridges to be carried by an infantryman in combat. The maximum effective range and muzzle velocity is about the same for both rifles at 1,500 feet maximum effective range and 2,800 feet per second muzzle velocity. Both weapons were, and are, among the finest military rifles in the world in my opinion.

After our issue of rifle, bayonet, and bucket, we were marched back to the barracks where we stowed away our bucket issue and bayonet. We carried our rifles to noon chow and stacked them in vertical stacks via the stacking swivel for the first time. Luckily no one's rifle hit the deck while stacking them otherwise we would have had to sleep with our rifle.

After chow we hit the drill field with our rifles for the first time. We spent the rest of the afternoon learning to coordinate the cadence calls with the manual of arms while trying to keep in step. By 1700 hours we had the basics down and could do a fair imitation of a platoon that knew what they were doing.

With our first time at drill with our rifles finally completed we headed for the physical training area and were again introduced to the three mile run. This time we had our rifles to carry which gave us eleven more pounds of weight to tote around the large field as we ran. We had to help seven or eight recruits cross the finish line that evening. The field jackets we were wearing became a burden and a hindrance to us while running but it was still too cold to not wear them. I was beginning to dread the coming days when we would be required to not only run with our rifles but to also have a 30-35 lb. field pack strapped to our backs. I was trying hard to just take it one day at a time.

That night we entered the squad bay to find our footlockers and bedding once again on the deck and in total disarray. The cleanup seemed extra tough because of our fatigue after running those six miles today. The stress, long hours, and physical conditioning were beginning to take a toll on us. Several men had already reported for sick call that morning. I was feeling sick and miserable and I wondered how long it would be before I had to report for sick call. The possibility of being removed from training because of sickness kept lots of recruits from reporting small injuries. Several recruits already had slight injuries and were hiding them from the Drill Instructors.

That evening we had our first mail call and the first chance to write a letter home. Free time was allowed at 2130 hours and I wrote Wanda a letter and included notes to Mom and Dad and to my Granny in the same envelope. I didn't mince words and freely let them know how tough Parris Island was. I highly encouraged them to send me all the news as I was very anxious to hear from home. I hadn't yet received a letter and was very homesick. Very few recruits had received mail that night. Mail hadn't caught up with all of us yet.

I was so stressed out when I wrote home that evening that my hand writing on the envelopes was twice as large as I would normally have written. I had to write the letters in such haste that no attention was paid to spelling or any such small detail as legibility. The main thing was to hurry and get a letter in the mail. The lucky few that had received letters sat on their buckets we had been issued earlier that day and grinned as they were reading. I envied them and was happy for them.

The buckets had a use other than for carrying water. They were used as our "chairs" during boot camp. We turned them upside down during free time and would sit there writing our letters on our knees. We would also clean our rifles and sew any tears in our utilities while sitting on the buckets. No other sitting was allowed at any time in boot camp unless it was at chow, at church, or during classes. At most other times we were standing at the position of attention.

That night we also learned to secure our rifles in the gun racks which sat in the middle of the squad bay deck. We used small chains to chain them to the rack and combination locks to lock them up. We had memorized our rifle numbers almost immediately after they had been issued that day. Any time A Drill Instructor or an inspecting officer asked your rifle number it was advisable to know it. That is, unless you were a glutton for punishment.

As we jumped into our racks that night and were lying at attention, the Drill Instructors led us in the "Rifleman's Creed" which every Marine is expected to know by heart. The words were painted on the large window of the Drill Instructor's office which we could see from our racks. That window was also used by the Drill Instructors to keep watch over us "Maggots" and "Scumbags". As we lay there we recited the words as follows;

This is my rifle.   There are many like it, but this one is mine.

My rifle is my best friend. It is my life. I must master it as I must master my life.

My rifle, without me, is useless. Without my rifle I am useless. I must fire my rifle true. I must shoot straighter than my enemy who is trying to kill me. I must shoot him before he shoots me. **I Will.**

My rifle and myself know that what counts in this war is not the rounds we fire, the noise of our burst, nor the smoke we make. We know that it is the hits that count. **WE WILL HIT...**

My rifle is human even as I, because it is my life. Thus, I will learn it as a brother. I will learn its weaknesses, its strength, its parts, its accessories, its sights and its barrel. I will ever guard it against the ravages of weather and damage as I will ever guard my legs, my arms, my eyes and my heart against damage. I will keep my rifle clean and ready. We will become part of each other. **WE WILL...**

Before God, I swear this creed. My rifle and myself are the defenders of my country. We are the masters of our enemy. **WE ARE THE SAVIORS OF MY LIFE.**

So be it, until victory is America's and there is no enemy, but peace!

We would repeat this creed every night as we lay in our racks at attention. This reminded us of the reason we were all there. The importance of understanding that our training as Marine riflemen was an absolute requirement for the protection of the rights and liberties of the citizens of the United States was made clear by this creed. It was a powerful motivation tool for the recruits in training on Parris Island, especially during a time of war.

By the end of this first week we had our routine worked out as to what would take place each day. Reveille was sounded at 0300-0400 hours. The recruits then attended to personal hygiene and clean up of the barracks. Chow was at 0530 then we ran the morning three mile run. Classes on first aid, military customs, military courtesies, regulations, etc., would be presented between frequent bouts of drill. After the noon meal there was more of the same until evening chow at 1700 hours. The three mile run was worked in again either before the evening meal or a couple of hours afterward. Mail call and "Square Away Time" was usually from 2130 to 2200 hours. On Sundays we attended church services. Calisthenics would also be included frequently along with more drill.

Chapter Nine

ORGANIZATION AND STRUCTURE

Learning the Ropes

One of the 3rd Battalion Recruit Barracks, Parris Island, South Carolina March, 1966. (USMC Photo.) Platoon 381 was located on the second deck of the barracks. Each barracks held a series of Recruit Training Platoons. The series I trained with was Platoons 380, (381-my platoon,) and 382. The peak training month (for number of recruits) was the month I started my training, March 1966. A peak training load of 10,979 recruits was reached during that month. The average number of recruits attending boot camp today is about 19,000 annually.

Our platoon continued to lose at least one recruit daily, on average, due to medical reasons, training incidents, or other reasons. We were down about a dozen recruits by the end of the first week. Our barracks had about thirty racks on each side of the squad bay. Each of the racks had a bottom and a top rack, which meant a total of one hundred and twenty sleeping spaces were available for the recruits. We had several empty racks due to the dropouts but every platoon always picked up some recruits that had been dropped from earlier platoons. By the end of our second training week we had picked up twenty three of these dropouts which gave our platoon nearly a full complement of recruits.

At this point in time the Drill Instructors had begun to assign a select few recruits to positions of responsibility and leadership. I wasn't among the few selected but I had never expected to be, as I had been feeling so poorly that just surviving day by day was my main goal. There were seven men chosen but as each one would screw up he would be replaced by another recruit to test his mettle in the position. The seven positions of responsibility included four squad leaders to head one of each of the four squads, a platoon guide to carry the platoon's guidon (flag), a "house mouse" that cleans the Drill Instructors' offices, and a scribe, who was responsible for maintaining administrative records. The only two that kept their jobs for the whole training period was the scribe and the house mouse. The poor squad leaders were changed just about every day when one of their squad members would screw up in some manner. It took at least four weeks to find four squad leaders that managed to keep their jobs until we graduated boot camp.

Sometimes a squad leader was dethroned when he was challenged by another recruit for the job. This challenge was often settled by going to the head in the dead of night where the best man would emerge with a few bumps and bruises and the squad leader job. The de facto leader of our platoon, Private Stevens, emerged after a few such challenges on the strength of his ability to defend his job with his fists. Private Stevens also received the award for being the Honor Graduate of our platoon at the graduation ceremony.   All recruits in positions of authority also received promotions to Private First Class upon graduation from boot camp.

Saturday, April 2nd we tackled the Obstacle Course for the first time. The course had about 20-25 difficult obstacles laid out in a figure 8 shape. Recruits were given a set time to complete the course. The time limit was around five minutes or less. There were many obstacles that involved logs, such as climbing up and over logs, vaulting log obstacles, and climbing a log wall. There are steel pipes, ropes and overhead bars that make up some other obstacles. All events are very strenuous and leave every recruit gasping for breath at the finish line the first time he attempts the course.

After I completed the course that first time I knew I had a lot of work to do before I could complete the course in the allotted time. Only a handful of our platoon managed to "beat the clock." As punishment the whole platoon had to go through it again. The second time around no one managed to come in on time. Before we left the Island every member of our platoon could complete the course in the time allotted with very little difficulty. We were slowly making improvements to our stamina and conditioning as each day passed. Our drill and manual of arms were vastly improved after that first week also.

I was still feeling poorly physically and was eating the barest minimum possible at every chow period due to feeling nauseous when I would see, taste and smell the food being served. The food at 3rd Battalion mess hall was excellent and I had no complaints about the quality or the quantity of the food. The continuous nausea was probably due to the stress and home sickness I was suffering from. Of course many others were suffering the same symptoms as I, but I had always been a picky eater which made my condition worse, in my opinion.

After our introduction to the obstacle course, and our having to execute it twice, I was feeling sicker than ever. My stomach was in knots as we lined up for chow at the mess hall and went through our "step, face, uncover, move," routine. As I went through the serving line I really felt that I wouldn't be able to eat anything at all, but I knew better than to go through the line and gather around one of the tables with an empty tray. I decided to take some spaghetti, a piece of sliced bread, and some lettuce salad only. I thought I might be able to force my stomach to accept small portions of those food items. The only problem was that the messman evidently tried to compensate for my mostly empty tray with a heaping measure of spaghetti and sauce. I knew I would never be able to eat that much food with the nausea factor I was feeling. The adage of "take what you want but eat all you take" was coursing through my head as the Drill Instructor shouted "Ready seats" and we sat down to eat.

I managed to eat a couple of bites of spaghetti before the nauseous feeling became overwhelming. I laid my fork in my tray and sat there at attention as I wondered if I was going to succeed in holding even that small amount of food down.

It was only a few seconds before one of our Drill Instructors noticed either my greenish face hue or the fact that I wasn't eating. He immediately proceeded to my side in a rush. He leaned over me as he put his face near my right ear and shouted, " What the f...'s your problem recruit? Clean your tray up! Hurry up!" As I sat on the bench at the position of attention I shouted, "Sir this recruit is feeling sick and can't eat all his food, Sir!"He shouted back at me, "I said eat it all! Eat it now, and that's an order!" He had attracted a fellow Drill Instructor's attention with his shouting and that Drill Instructor also rushed over to my side and joined in as they both proceeded to shout and gesture with unintelligible gibberish into my ears. I managed to understand the gist of what they were shouting. I could either eat everything in my tray or wish I had.

I was already having a hard enough time just keeping my position of attention without falling over, much less trying to eat a super helping of spaghetti. Nevertheless I had no choice but to try to humor my antagonists. I picked up the fork and forced a bite of spaghetti into my "green around the gills" mouth. As soon as it hit my stomach I felt for sure it was all coming back up. "Eat it all, maggot!" was still ringing in my ears as I tried to take another bite but a wave a nausea came over me just then that set me to gagging like a dog chocked on a bone. By that time my fellow recruits at the table probably thought sure that I was going to upchuck but somehow I held it down as the gagging continued.

When the two Drill Instructors saw me struggling with my dilemma, they immediately stopped their shouting and both withdrew from my side. They realized by then that my refusal to eat wasn't an act of defiance but was caused by genuine sickness. I wasn't punished when I carried my tray to the two garbage cans and raked all the food off into the can.

We recruits didn't yet know it but when a recruit showed tenacity and courage in the face of adversity the Drill Instructors respected them and would ease up a little. When they saw how sick I was and that I had trained all day without complaint they cut me a break. Thankfully, after chow we had classes instead of physical training or drill the rest of the evening. The obstacle course had taken the place of the three mile run for that day and that was also in my favor that evening. I probably wouldn't have been able to run three yards, much less three miles, as sick as I was.

The next morning I felt much better and even managed to eat a light breakfast. I again attended 0900 church services in the 3rd Battalion chapel. Going to church helped our morale and gave the recruits the spiritual help they needed to cope with the stress of boot camp. I would always pull out my small New Testament each night during our square away time and read a few verses before lights out. Coping with adversity the recruits had to face daily would have been nearly impossible without spiritual help, in my opinion. I do know it made a big difference in my ability to stay the course.

I always enjoyed the classes on Marine Corps history and customs very much. The classes are for motivational purposes as much as they are used to teach recruits about the tradition, leadership, and professionalism of the Corps. The Marine Corps heroes that the many campaigns and battles had produced since the Corps initial forming at Tun Tavern in Philadelphia on November 10, 1776,are brought to life by these classes.

We learned about the exploits of Dan Daly, Samuel Nicholas, Lewis B."Chesty" Puller, and many others who had excelled as leaders and heroes in days gone by. We also studied the many battles the Marine have participated in over the years.

We also learned about the three core values of the Marine Corps. The character of a Marine is defined by these three values which are really the foundation of the Corps.

Honor: Each Marine must exemplify ethical and moral conduct.

Courage: The moral strength and the will to heed the conscience to do what is right

Commitment : Total dedication to Corps and country.

We had, by the end of the first week, been required to know our eleven general orders and be able to rattle them off to any commissioned or non-commissioned officer that asked us to recite them. Monday morning after our three mile run and more drill, we had a class on Interior Guard. This class basically dealt with what was expected of us when we walked a post on guard duty outside the barracks and when we stood fire watch in the barracks. Fire watch was stood in the barracks by several recruits every night as the other recruits slept.

Smoking was not allowed during training for the first few weeks but that didn't stop a few recruits from somehow acquiring cigarettes and a lighter and sneaking a cigarette in the head after lights out. The fire watch was supposed to report these violators but rarely did.

Our Senior Drill Instructor made the comment one time that, " I can see the barracks from my house and last night I saw so many lights in the windows of the head that I thought we had a fire in the head. I almost called the base fire department!" That was his way of warning the recruits that nothing we did went unnoticed.

After three weeks the recruits that smoked were allowed to smoke in a group, but only one cigarette at a time. The Duty Drill Instructor would march the smokers outside and "light the smoking lamp" for one cigarette and one cigarette only. Sometimes he would allow them one or two puffs before he would shout, "The smoking lamp is out! Put'em out!" If one puff was taken after his admonition all smoking privileges were suspended for an indefinite period.

After watching the smokers have the privilege of getting to smoke their one cigarette per day I was feeling a little left out. The non-smokers had to stay inside and clean the barracks while they were puffing away and I began to resent it. One evening my anger got the best of me and I threw my cleaning rag into my bucket and marched to the Drill Instructor's office and stood beside his hatch. I assumed the position of attention and rapped sharply on the side of the hatch with one arm swinging over my head while my other arm was at my side with my thumb alongside the seam of my trousers.

I slapped the frame of the door three times as per requirement and the Drill Instructor shouted, "Speak Dipshit!" I was supposed to then take one step into the center of the doorway, face left, and scream loudly, "SIR, Private Nickels requests permission to enter, SIR!" Then he would give me permission to enter providing I hadn't screwed up. Instead, when he shouted, "Speak Dipshit," I rushed into his office before I spoke. I stood at attention in front of his desk and screamed loudly," SIR, the Private would like to request that the non-smokers be allowed to purchase and eat some "pogey bait" (candy) since the smokers get their smokes, Sir!" There it was. I couldn't take it back. He would either throttle me or allow the non-smokers to have their pogey bait. My dander was up.

The look he gave me was indescribable. I could see the veins in his neck pulsating as he stared at me as if he couldn't make up his mind what to do with me. Now *his* dander was up. Finally, he shouted, " Private there was no permission given to enter my office. Give me five hundred side-straddle hops and count off each one as they're done!" I was still side- straddle hopping long after the smokers came back inside. I regretted my sugar craving attack. I never did get any pogey bait all through boot camp. They're probably still laughing at me today.

We also were not allowed to have any cakes candy, or cigarettes sent to us by mail. One recruit received one flattened cigarette in a letter and it was discovered by a Drill Instructor during mail call. The recruit had to chew it up and swallow it. My wife sent me a wedding band in a letter and the Drill Instructor had me open it in front of the platoon and put it on. He then stood me at attention twenty minutes or so as he bantered me with such phrases as, "The little wife is afraid her boy might stray while he's here at Parris Island so she sent him a golden ring to wear on his pinkie! Bless its little heart!" I had never smoked but I wished she had sent me a flattened cigarette in the envelope to eat just then instead of the ring.

The second week of training we received our military identification card and dog tags. My dog tag had the wrong blood type listed on it. I never was able to get it straightened out. I was required to wear my dog tags at all times. I was fortunate that I never needed blood. My blood type is actually O Negative instead of type B that's listed on the tags.

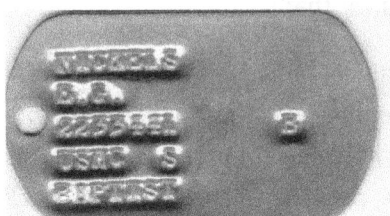

My identification photo looks more like a mug shot than a U.S. Marine Corps I.D. photo. My photo shows the stress and lack of sleep I was clearly suffering from.

Our first training in Marine Corps Martial Arts occurred near the end of this second training week. The training given the recruits at that period of time was Judo. Judo concentrates on using the weight and actions of your opponent to stop his attack on you. We gathered in a large pit where we received instruction in how to disarm and even kill an armed opponent.

We took turns practicing the different moves on each other until we had absorbed enough basic knowledge to protect ourselves in close combat situations. We practiced in two man teams and my opponent tried to get me to agree to take it easy on each other but I didn't agree to do so. We had been warned that if caught taking it easy on your opponent we would be shown by the martial arts experts how easily they could injure a man. I didn't want to take that chance. So we ended up slamming each all over that pit. I was so sore the next day and had so many bruises that I considered going on sick call. We had judo practice nearly every day after that.

Each night after our day's training ended we would sit on our galvanized buckets and disassemble and thoroughly clean our M14 rifles. Rifle inspection was held each morning after chow at around 0630.There was also a personal inspection held each night between 2100 and 2200 as we stood at attention in front of our racks.

Each and every night , including Sunday night, we would turn the squad bay lights out and disassemble our M14's and reassemble them in the dark. We were timed in this exercise and were given extra physical training if everyone didn't have

119

their rifles reassembled when the lights were turned back on. Eventually we could literally disassemble and reassemble our M 14s blindfolded. We were becoming more comfortable with the adage that our rifles would be like an extension of our natural selves during our training and afterwards.

We began bayonet training in the latter part of this week. We practiced the vertical butt stroke, the horizontal butt stroke, the slash and parry, the jab, etc. We paired off with each other, and with our scabbards covering our bayonets we practiced parrying bayonet thrusts an enemy might confront us with.

The majority of our bayonet exercises that day and nearly every day there after concentrated on mostly offensive moves with the bayonet, not defensive. Any training we had at Parris Island was geared towards offense only. We were told that as Marines we would always be moving forward, not waiting for the enemy to attack us. (In Vietnam Marines weren't allowed to fight as they had been trained. The frustration of Marines during that period was at a high level because of not being allowed to go on the offense as a means of taking the war to the enemy. We were destined to set up enclaves and wait to see what happened. That's no way to fight a war, in my opinion.)

One exercise of our bayonet practice had us attacking large dummies that were hanging from wooden beams on the practice field. We then attacked the dummies "en masse" as we screamed our lungs out and yelled "Kill' kill, kill!" as we made our bayonet slashes. I have to admit that I was pretty motivated as I attacked those dummies. All I had to do was imagine that each dummy represented a Drill Instructor. Having been through these same training procedures themselves when they attended boot camp, they were aware of our thoughts as they had once had them themselves. They would even urge us to be aggressive by shouting, "Imagine that dummy is your worst nightmare of a Drill Instructor and you now have a chance to get even with him!" With those words I shouted as loudly as a Confederate soldier yelling the "Rebel Yell" at Gettysburg when Pickett was making his charge against the Union Army! Those Drill Instructors really know how to get you motivated.

On Saturday evening April 9th we had our first introduction to the Confidence Course. The Confidence Course is geared to not only test your physical ability, but to challenge the recruits psychologically with more intricate obstacles than the Obstacle Course you had already tackled. Many of the obstacles deal with height and are designed to challenge recruits to climb heights without fear.

There are fifteen to twenty obstacles such as the infamous "Slide for Life" where a recruit climbs a high tower and has to climb from the safety of the tower onto a rope that stretches from the tower horizontally at an angle to the ground. The rope goes over a large pit filled with water. About two thirds of the way down the rope, (providing you haven't fallen yet,) you have to turn completely around on the rope, swing your legs back upon the rope and slide on over the water.

I managed to complete this exercise successfully the first time without falling into the water, which we had been told was full of alligators, although I didn't see one. Of course this was a highly motivating factor for me that day. Several dozen of the recruits fell into the water and were forced to contend with their wet utilities until they dried with the help of the hot sun.

Some other obstacles were named the Sky Scraper, the Belly Buster, and the Tough One. Climbing a log tower by grasping rounded logs as you hoist yourself from level to level thirty to fifty feet into space will get your senses to a higher level I assure you.

Most recruits would more than likely dread the Confidence Course more than the Obstacle Course but I was the exact opposite. I had never had a fear of heights but the Obstacle Course required brute arm and leg strength to conquer, which was my weak point. Any exercise that required arm or leg strength was a challenge to me during boot camp. The exercises that required running, marching, classroom study, or confidence to complete, I was always able to do very well in. As boot camp progressed, so did the body strength of most of us. By the sixth or seventh week of training I discovered that I could handle the obstacle Course with ease and even went from struggling with ten pushups in the first week to being able to do fifty or more by the fourth week.

The day after we ran the Confidence Course was Easter Sunday, April 10, 1966.After chow the Protestants among us were marched to a parade ground where Easter services were to be held. Although we had on our field jackets the wind coming off the ocean cut through our jackets as if we were naked. I was colder and more miserable that Easter Sunday than I had ever been or have been since. I literally thought that my ears, hands, and feet might be frostbitten by the time services ended one and one half hours later. One would think that an island in the southern hemisphere would be experiencing too much heat in April instead of frigid cold, but not that year at least. It was the end of April or later before we experienced really hot weather.

That evening we were introduced to a new exercise we hadn't practiced before. Just before we were allowed our few minutes of square away time we were told to drop to the deck on our knees. We were still required to stand at attention in front of our racks all the time unless engaged in an exercise or cleaning the barracks, so we always looked forward to being allowed to sit on our galvanized buckets at night. When we were told to drop to our knees at least we weren't at attention and that was relief in itself.

As we dropped to the concrete deck the Drill Instructor said, " When I give the word I want every swinging d... of you to slap the palms of your hands against the deck and there better not be any slacking off! We're going to toughen those delicate little lady hands up until we're able to break a rifle stock or two each while you're here on Parris Island. When I give the command of left "shoulder arms" I want to hear the sounds of those stocks cracking and shattering! Now to prepare for breaking those stocks we have to toughen those dainty claws of you p...... whipped babies. Now get to smashing those precious girlish palms against that deck!"

After thirty minutes or so we were brought to attention with our beet red and aching hands along the seams of our trousers. I felt like a million bees were busily stinging my hands for a couple of hours afterwards.

The Drill Instructor wasn't blowing smoke with the hand exercise, as there were many rifle stocks shattered in boot camp when going from a "right shoulder arms" to "left shoulders arms." Almost every night we did some serious time with slapping our hands against that concrete deck. Try as I might (and I did try very hard) to crack the stock of my rifle throughout the many hours of drill on Parris Island, I was never able to do it. The only thing I managed to do was to strain my biceps in trying to do so.

As we marched along and were engaged in doing the manual of arms we would hear a loud, sharp "craccckkkkk" as a recruit would manage to shatter his M14 stock. The Drill Instructor would then halt the platoon and call the recruit to the front of the platoon to hold his shattered rifle up with the butt stock dangling by the sling. The Drill Instructor would make a big deal of telling the recruit his "free time" would commence as soon as we entered the barracks at 2000 hours. That would mean an extra hour of relief from standing at attention.

I have often seen movies where recruits in training would relax and go to town to tie one on during training, but that is Hollywood's version of training. We had no newspapers, no TV, no radio, no candy, pop, no free time to even go outside the barracks in the evenings or on weekends. Saturday was just another training day and Sundays were for church, doing laundry, cleaning our rifles, gear, and the barracks. In some of my letters home I would say, "I still haven't had a drink of pop here!" That was in reference to the fact that I grew up not drinking anything but pop throughout my young life up until then. Everybody in the family knew my preference for drinking pop. I hardly knew what water tasted like at home until I was forced to drink water on Parris Island.

Chapter 10

MAKING THE GRADE

Drown-Proofing and Testing

Attacking the" Dummies" at Bayonet Drill. USMC PHOTO

The third week of our training began with academic testing. We had spent hours and hours of classroom lectures learning about first aid, Marine Corps history, the Uniform Code of Military Justice, and all about the nomenclature and capabilities of the M14 rifle, among other subjects. This week would allow us to test our knowledge of what we had learned so far.

Failure is not an option during the examination on these subjects and much preparation is made in practicing to pass the exams. Our Drill Instructors were very concerned that a few of the recruits might not make the grade needed to pass. The final test consisted of around eighty multiple questions and a total of two hours were allotted to finish the exam. The questions were moderately difficult and were easily passable providing you had paid attention in class. Paying attention wasn't always possible when you were getting only six hours of sleep and training in a near brutal fashion an average of twelve to fifteen hours a day. The temptation to close your eyes and catch a wink or two while listening to the Instructor in his droning voice was tempered by fear of the consequences if caught. Some recruits managed to do it anyway. The Drill Instructors' concerns were well founded.

The passing grade was achieved with answering sixty or more of the questions correctly. I personally thought they were quite easy but my interest in history and the ability to memorize the main points of a lecture stood me in good stead for this part of boot camp.

There was one recruit in particular that the Drill Instructors were worried about passing the exam. This recruit had been one of our many pickups from having been set back in training from an earlier platoon. We had lost around twenty five of our original number by now but we had accrued about that many more from one or more of the special training platoons, such as the Medical Hold Platoon or the Physical Conditioning Platoon, or even from the Correctional Custody Platoon where a recruit with minor disciplinary actions are sent.

This particular recruit received several hours of one-on-one instruction from the Drill Instructors as they tried to help him pass this first academic goal of boot camp. Incredibly this was his twenty sixth week of boot camp and they wanted to make sure he succeeded in all his training requirements. I had heard that twenty six weeks was the limit for boot camp recruits and if one wasn't able to successfully complete the requirements by that time you would receive a General Discharge.

After completing the tests we proceeded to the drill field and drilled, drilled, drilled, and drilled some more that day. We were finally getting to the point where mistakes were few and we were looking more like a "squared away" platoon of Marines, although none of our Drill Instructors would ever admit as much. We were certainly never called "Marines" in boot camp by anyone. That's a title earned only on the last day when the Eagle, Globe, and Anchor is awarded and recruit graduation is held on the parade ground. A recruit is subject to being set back, discharged, or dropped to a later arriving platoon at any time during training if he isn't able to make the grade expected of all Boots.

On Tuesday morning the Drill Instructors were informed that 100% of our platoon had passed the Academic Evaluation Test. Our Drill Instructors were ecstatic with the results as the results of any evaluations the platoon faced reflected back to them in the eyes of their superiors. It was the first time I remembered seeing a smile on any of them. That same evening we were rewarded as a platoon by having a large yellow achievement ribbon or banner tied to out platoon guidon by the Drill Instructors. Each time a platoon has an achievement that is tops in the series of three platoons they would be awarded a different colored banner to signify their top placement. It would be hard to describe the pride our platoon felt with the awarding of that banner.

Every platoon at Parris Island took pride in their guidon that the Platoon Guide carried when the platoon was marching. The flag carried the number of their platoon for everyone to see and we were proud of the 381 that emblazoned in large gold letters on our red colored guidon.

Pride in each individual platoon could sometimes lead to some awkward situations. Once our platoon and another platoon from the 2nd Battalion met at a crossroads and our platoon was crowded out of the street by the other platoon. Our Drill Instructors were livid at the timidity of the head of our platoon by not holding their ground and "plowing through" the 2nd Battalion platoon. They double timed us back to our 3rd Battalion area which was over a mile away and gave us an extra dose of PT that evening for good measure. Before hitting the rack they called a school circle to inform us that we were never again to allow another platoon to take precedence over our own platoon, especially while marching.

We took their warning to heart. The next time a strange platoon tried to advance into our territory we marched through them and left their formation a shambles. We had learned our lesson well.

The Parris Island training areas and drill fields were especially crowded in March, April, and May of 1966, which caused a lot of clashes between platoons. The peak training month in all the Vietnam War period was March, 1966 when 10,979 recruits were in training at Parris Island. This was the same month that my platoon (381) started their training. This represented the highest monthly total of recruits training at the same time since March, 1952 when a peak of 24,424 Marine recruits was trained for the Korean War that month.[1]

In this third week there was a noticeable reduction in the stress level in the mess hall and in the barracks. There was no reduction in the level of physical training but there was not as many one- on- one stressful incidents. Our four Drill Instructors were reduced to two during the day and only one at night. They had learned by now that they could trust the remaining recruits to do their duty in training since they had weeded out over two dozen of the original group by this time.

Our platoon had lost one Drill Instructor the second week without explanation but it was probably for the best. The one removed was the exact same one that had "put me to sleep" that first Saturday as we met our Drill Instructors for the first time. His aggressive behavior went way beyond the rest of the other Drill Instructor's actions and I have always believed this was the reason for his removal. I wasn't the only one to experience his ire that day and I'm sure the Senior Drill Instructor was aware of everything going on.

The other three Drill Instructors were tough but fair as we continued our training. In the last week of our First Phase

---

[1] http://www.globalsecurity.org/military/facility/mcrd-parris-island.htm

Training a new Drill Instructor was assigned to our platoon to replace the "Choker". His name was Sergeant M. Garcia. Like our three original Instructors he turned out to be a fair minded Drill Instructor. His special mode of punishment for a wayward recruit was to use four fingers of one hand to prod your stomach when he felt you were out of line. I had imprints of his fingers around my belly button nearly every day after he joined us. It was nearly impossible not to screw up in some small way to deserve punishment (in the Drill Instructor's eyes) nearly every day. The other Drill Instructors had their own unique system of punishment, most of which involved extra PT on the quarterdeck. [2]

Our Senior Drill Instructor was Staff Sergeant T.T. Lawson. He was a small man whose toughness belied his small stature. As the Senior Drill Instructor it was his job to keep the other Drill Instructors straight and to make sure they kept us recruits straight. He did a fantastic job and I always had a lot of respect for him.

The other two Drill Instructors were Sergeant J.C. Shue and Corporal M.W. Meade. They, along with Sergeant Garcia were responsible for the day to day training of our platoon. I had a lot of respect for the job they did also.

Lt. Colonel G.L. Lillich was the 3rd Battalion Commander, Captain H. L. Haley was our "S" Company Commander, 1st Lieutenant E.E. Record the Series Officer, Master Sergeant H. Massey was the Chief Drill Instructor for the Series, and Gunnery Sergeant D.J. Kane was the Series Gunnery Sergeant. These Marines were responsible for the training of our series of three platoons of recruits and all did a respectable job of

---

[2] Area outside the Drill Instructor office in the barracks.

molding our three platoons into qualified Marines upon graduation from boot camp.

Just when our platoon had started getting to where we were feeling as if we were getting good with our manual of arms and as if we could march and drill with the best of any platoon, the Drill Instructors threw a cog into our false sense of security. They began to call selected individual recruits into their house (office) after training each day and proceed to inform them that it still wasn't too late to be set back or dropped from training if their performance didn't improve considerably. After surviving what we would come to call "Hell Week" (the first weekend and first full week of training) we had begun to think that the remaining recruits were safe and secure from any further training drops.

The recruits whom had developed a little bit of an attitude were an especial target of the Drill Instructors' wrath. The fact that they were performing well in all other aspects of the training regimen didn't mean squat if a recruit had an attitude problem. I don't know if any recruits were set back in their training after their evening sessions with the Drill Instructors but the quarterdeck was well inhabited most evenings of this particular week.

I have always thought that the Drill Instructors sensed our sense of security and wanted to relieve us of any thought that they were satisfied with our performance. No matter how well we did in any event the expectation was that you could do better if you put 110% of your effort into it.

Whatever the reason for the increased scrutiny of certain recruits it worked. We all picked it up a notch in training and the competition to be the best in every training event was fierce throughout the remainder of boot camp.

The Drill Instructors selected several recruits to spend two hours each evening in a special training platoon called Motivation Platoon. This platoon was normally one to which those recruits that lacked proper interest or motivation were sent for several weeks before beginning their training. Those that were sent from our platoon were allowed to continue training with our platoon while marching to the Motivation Platoon each evening and marching back to the barracks just before lights out. They seemed to enjoy the extra training and were even allowed to fly their own flag as they marched, which was emblazoned with the words, "Tiny Tim's Recon Gang."They took it in good stride and with good humor.

Our platoon spent a lot of time drilling and practicing the manual of arms as this was the week our series of three platoons would meet in competition with each other in an event called Initial Drill Competition. This event would showcase how well each platoon in the series was doing in various drill evolutions. The top platoon would get a trophy and a streamer to attach to their guidon for their efforts.

There was always furious competition among all platoons on Parris Island, especially those in the same series and company. To be number one is the goal of every member of every platoon and this event topped them all. After all, marching and drilling was one of the things that any Marine always took pride in. The sound of three or four hundred rifles ringing with the sound of wood being slapped as the manual of arms is being performed is a beautiful sound. We were all anxious to show what we could do that Friday morning.

We marched to the parade ground early in the morning where our three platoons, 380,381, and 382, observed Morning Colors. Afterwards, Series Chief Drill Instructor Master Sergeant Massey commanded platoon 381's Senior Drill Instructor, Staff Sergeant Lawson, to put his platoon through its drill evolutions .

S/Sgt. Lawson was handed a list by M/Sgt. Massey of up to forty evolutions to put our platoon 381 through as a group of officers and non-commissioned officers observed, including our Battalion Commander and Company Commander.

There was tension in the air as S/Sgt Lawson ordered us to "Fall in!" He then gave us the command of "Forward, March!" As we marched along the parade field he first ran us through the simple evolutions of drill such as, "Left, face! Right face! "To the rear, march!" We also did the manual of arms as we were marching, such as "Port arms! Order arms! "Inspection arms!"

Some of the marching evolutions are more difficult than others, such as marching by file and to the rear which causes the rear rank to have to march in the lead, which is not very familiar to them. These several difficult evolutions are usually when you see some screw ups such as skipping and hopping or marching out of step. When going from left shoulder to right shoulder or right to left shoulder while doing the manual of arms it's easy to set the rifle on your shoulder at an angle instead of pointing straight. These mistakes of alignment are duly noted on clipboards carried by Senior Drill Instructors from the other two platoons. They missed nothing and were meticulous in their observations and grading of our platoon as we did the required evolutions.

After the twenty minutes allotted for the various evolutions of our platoon, the next platoon, platoon 382, went through their evolutions while the senior Drill Instructor of our platoon and platoon 380 observed and noted each discrepancy committed during the performance of the evolutions of drill.

Platoon 380 had performed first so after 382 finished, we had a few minutes of waiting as the judges added up the grade. Each member of our platoon knew how important to our four Drill Instructors this decision was. It was probably more important to them than even to the recruits, as the results would be seen as a direct reflection of how well the Drill Instructors had trained their charges.

After a few minutes the three platoons were called to attention and the Company Commander, Captain Haley, stepped forward to award the trophy for the best drilled platoon in the Series. The Senior Drill Instructor of each platoon stood at attention in front of their platoon and waited as Captain Haley marched to the front of the formation and awarded the trophy to a platoon other than ours! I'm not sure which platoon was awarded the trophy but they were ecstatic.

Each recruit in our platoon standing in that formation thought we had been robbed of our trophy. Of course the other losing platoon thought they had also been robbed. The Captain congratulated all three platoons for their effort and told the formation that the competition had been razor close. Our platoon had come in second but that was poor consolation to soothe our wounded pride.

The Drill Instructors took the loss better than the members of our platoon expected, with the only punishment being of the verbal sort with some intense drill sessions in the following weeks.

That night I received a letter from Wanda that she had decided to give our house up and was selling our furniture as she had no place to store it. She was going to stay with my Dad and Mom until we found out where the Marine Corps would send me after training. We could never make a decision as to what she should do about keeping our home together, so I was glad she had decided on her own. Actually it wasn't a hard choice because we both knew I wouldn't be making enough to keep the bills paid. Wanda was working part time which helped very little. I hadn't received any pay yet and wouldn't until my graduation from Boot Camp. I had applied for an allotment for Wanda during my second week in boot camp but she also wouldn't receive a check until around the time I graduated boot camp. The allotment would be for around $60.00 with a part of it being taken out of my pay. My salary as a Private E-1 was $87.90 per month and I owed the Marine Corps about $ 26.00 of that first check due me!

Our platoon was marched to the PX in our third week and we were given baskets as we were marched through the PX and told what to put into our baskets. The bulk of our purchases were personal items such as shaving cream, tooth paste, foot powders, rifle cleaning patches, soap, razor blades, etc. The total was about $ 26.00 which, as I already mentioned, would be subtracted from our first check.

I probably could have done without the razor blades as I needed to shave only because it was required. Shaving every day was new to me as I was at the age when once or twice a month was more than sufficient. I could have probably gotten by with one blade the whole time in boot camp. I could have put my razor blade money to better use by buying myself a bar of "pogey bait" instead.

One incident will attest to the above. We were having a Company Commander's Inspection in the company street with the sun shining brightly in our faces as the Captain looked closely at my face and said," Private, I would say that you missed a spot when shaving but all I see is fuzz and it's understandable how you might have missed it." It was unclear to me at the time whether he was complementing me for just leaving one hair or criticizing me for leaving one.

We went on our second conditioning march Saturday morning with full field pack including haversack, knapsack, and our rolled up shelter half. The weather was beginning to get very warm as it was now April 16 and Parris Island had begun to heat up. We hiked at least seven miles and it might have been closer to ten miles. I just put one foot in front of another and tried not to think about the miles. The conditioning hike of last week had been only five miles or so. I enjoyed getting away from the hustle and bustle of the battalion area and into the boondocks for a little while on these hikes but the heat, sweat, and the famous sand fleas were beginning to do their thing in the heat of the day.

We weren't allowed to swat the sand fleas at any time no matter how many were crawling in our mouth and ears. The only relief we could get was when the Drill Instructor had his back turned with his attention somewhere else for a moment when we would attempt to slap at the sand fleas without making any slapping sounds. It was almost comical to watch one hundred or more men take a swipe at the sand fleas almost as if it was a synchronized part of drill. If one small slap was heard or one unusual movement was detected by the alert Drill Instructor you would be sure to have to do PT right then and there until you nearly dropped.

Occasionally, after being driven nearly to a stage of frenzy by being bitten by the unremitting insects all day long, the Drill Instructor would allow us one slap in unison to try for some relief. When I said one slap that was exactly what we would get, one slap only. I would have them in my nose, my eyes, my mouth, my ears, and crawling on my face when the Drill Instructor would shout, " Ready! Hit'em!" You had better be ready when he gave the command because there were no second chances.

A few times when a recruit was caught slapping a flea we would be instructed to hold a mock funeral for it with the offending recruit digging a deep hole with his entrenching tool to bury the dead insect . Sometimes we would have to search for the dead flea that had been "murdered " by the recruit. After thirty minutes of searching I have seen a Drill Instructor look at a flea that an enterprising recruit has "found" and exclaim, "That's not the one you murdered. Keep looking until you heartless killers find the right one!" We learned to follow orders the hard way.

The famous sand fleas of Parris Island were every bit as bad or maybe even worse than I had been led to believe.

Water survival training or Drown Proofing as the recruits commonly called it, was a lesson in perseverance, especially for non-swimmers. This technique allows the swimmer to survive in deep water after their ship or boat sinks or after ditching an airplane in a deep water environment. This is possible by allowing the body to relax and to lie face down while floating. As you are floating you occasionally raise your head from the water and take a deep breath, then again let your head and body float until another breath is needed. With this method of water survival it is possible to survive for several hours while conserving energy.

Some people are natural floaters while others are sinkers. I was one of the sinkers. A sinker has a harder time with drown proofing because as he sinks he has to use more energy to rise for a breath of air. At least that was my experience.

Some recruits had a great fear of the water and had a very difficult time with the water survival training. Every recruit was required to attempt to stay in the deepest end of the Olympic size pool and float for at least one hour to pass the course. Several hours each day for the whole week were spent in learning how to do the procedure correctly and safely.

There was no time to teach every recruit how to swim before attempting to pass Drown Proofing. We were all required to attempt it. We had special help from the water survival instructors with our attempt to do this. The help they provided was mostly to take a long wooden pole and push the recruits that panicked back into the deep water. They were using the old sink or swim method.

As a child I had spent many hours in the local swimming hole known as the Bill Doug Hole but my swimming ability was limited at best. I attempted to Drown Proof but being a sinker and a non- swimmer to boot was too much to overcome. My years of dog paddling instead of learning to swim had caught up with me.

The recruits that couldn't drown proof were given another test for their second class water survival training which required the recruit to swim around the perimeter of the whole pool using any swimming method they chose. I started the course by swimming free style as far as I could, then switched to my old fall back dog paddling method. With my switch to my favorite swimming stroke I left the other swimmers in my wake. I had finally passed the part of boot camp that I had dreaded more than any other and now felt good about my chances to successfully graduate recruit training in a few weeks.

Our final water survival test was a cannonball dive from a thirty five foot high platform. Some recruits had a fear of heights and water which made for a difficult combination. Eventually every recruit successfully completed this hurdle of our training.

Chapter Eleven

MESS AND MAINTENANCE

Phase Two Begins

From left: S/Sgt. Lawson, Pvt. Walton, Pvt. Stevens, Pvt. Weaver—Platoon 381 in platoon formation.

With the beginning of week four of training we were entering our Second Phase. As Second Phase recruits we were extended the privilege of at last receiving permission to unbutton the top button of out utility blouses. This one small act meant more to the recruits than can readily be understood by anyone that hasn't had the experience of recruit training. Not only is the constantly tightly buttoned blouse an annoying impediment to normal breathing, especially when running a three or four mile conditioning run, but it identifies you as a First Phase recruit and a newly arrived Boot to other platoons.

There were three phases of recruit training at Parris Island during my time there in 1966. The first three weeks of training were known as the First Phase, the second three weeks were the Second Phase, and the last two weeks were the Third Phase. Each completed phase of training marks a successful milestone for the individual recruit. Each phase also marks an increasing degree of difficulty in the training regimen.

As we entered this fourth week we were not only expected to do our normal training duties, we were also expected to do either base maintenance duties or mess duty in our battalion mess hall. Every platoon in our series of platoons had a week of these extra duties during this fourth week. Our platoon got lucky and was assigned "maintenance and guard" which is a much better duty than mess duty.

Recruit messmen were roused from their racks at 0300 in order to help prepare for the morning chow call which began at 0430 hours each morning, except Sundays. Morning chow was at 0600 hours on Sundays. Arriving at the mess hall those recruits on mess duty had to change their green utility uniforms for white trousers and white skivvy shirts. (tee-shirts) Then they stood formation in the mess hall for a uniform and hygiene check before the serving of food began.

After every platoon has had their chow time, the messmen turn their attention to the clean up. Every chair, bench, and table is wiped clean, the decks are scrubbed and swabbed, and every part of the serving line is scrubbed meticulously.

The life of a messman was one that I didn't envy and I felt a large portion of relief when the Drill Instructors informed us on Monday morning April 18 that we would be performing maintenance duty all week in addition to our training duties. The eighteen hour days in the mess hall was something that I and my platoon mates could live without.

After morning chow we ran the Obstacle Course, had drill, and did PT for thirty minutes. Then we received our assignments for our first day of maintenance. I and several other recruits were assigned to march to a warehouse where we began unloading boxes of assorted cleaning supplies onto a truck from a box car shaped storage building.

We had a corporal in charge of our group that wasn't wearing a Drill Instructor's Campaign hat or a duty belt and didn't seem to have a Drill Instructor's demeanor. It wasn't long before we realized that he was going to allow us to even speak among ourselves like we were normal human beings for a change. We had never had this privilege in the nearly four weeks we had been on the Island and we felt liberated to a large degree.

For the first time we could carry on a conversation with each other without dire consequences resulting. As we worked we discussed among ourselves where each was from and how long each of us were in for. It reminded me of a gang of convicts having been in solitary confinement and being allowed on the yard for exercise for the first time in weeks. We jabbered as if there was no tomorrow and we had today only to express our feelings before our time was up.

The recruits I worked with that day were mostly from Boston, Massachusetts, with a couple from Pittsburgh, Pennsylvania, and two others from New York. I was the only Southern boy in the group and took some good natured ribbing because of my southern accent. One of the recruits said he knew I must be a Southerner when he first heard me pronounce "Aye, Aye Sir" as," I, I Sir!" I quickly told him that I couldn't help it if everyone else was pronouncing it wrong.

One thing the twelve of us working together that day had in common was the fact that we were all draftees. We ranged in age from eighteen to twenty four. Our education level varied from two college graduates to my recently acquired G.E.D. Three of us were married.

One of the college graduates was employed as a school teacher when he was drafted. His name was Private Moore from New York. As Private Moore and I discussed our situation while tossing boxes of cleaning materials I discovered that he had the same goal after boot camp that I did. That goal was simply to get our military obligation over with and get on with our lives. Neither of us had any pretention of wanting to make the military a career but at the same time we were proud to serve our country.

Another recruit I became acquainted with that day was Private Wilson, who was also from the state of New York. He was one of the three married recruits in the party. His wife was expecting the birth of their baby any day and he expressed to us that his stress level was at a severe level as he waited for word from home of a successful delivery. Like Private Moore and I, he also expressed the fact that he was anxious to return home after his obligated service was over.

After our discussion that day these two recruits and I became best friends throughout the remainder of boot camp. They were concerned enough about my skinny frame during the following weeks of training that they would pile their rarely served cookies and cakes on my tray at chow time in the mess hall. When I objected to this they simply stated they would expect the same from me if the shoe was on the other foot. They knew I rarely ate anything other than salads, bread, and potatoes. I was never a big meat eater and I very seldom ate more than a few bites of anything other than the desserts. They had noticed my sparse eating habits and were determined to help me get a few extra calories into my diet. Private Monte was also on a restricted diet as he struggled to lose a few extra pounds during training. On the other hand Private Watson was nearly as skinny as I was. The sacrifice of their desserts was probably instrumental in helping to keep me from losing even more weight during training.

AS we were engaged in loading supplies a Lance Corporal from Battalion Headquarters approached us and asked for a Marine recruit by name to come with him to Battalion Headquarters. When the recruit returned we asked him what the summons was about as it was highly unusual for a recruit to have to report to Battalion unless it was something of vital importance. He looked at us and grinned as he replied, "They asked me if I wanted a medical discharge because of my feet!" Remembering how badly we all wanted to go back home when we first arrived on the Island we all chimed in at once, " You told them you would accept it of course, didn't you?" "No, he said, I'm a third of the way through boot camp already. I've come too far to quit now!"

I couldn't believe my ears. Most of us would have accepted an honorable medical discharge in a heartbeat. I was certain that I would have accepted it if one had been offered due to my high blood pressure problem when I first arrived. I was anxious to find out his reasoning for his refusal for a discharge.

When we questioned him about the reason the offer was made he said he had a bad foot condition that made keeping up with the platoon difficult but that he was determined to keep on training if possible. He was a draftee but he was beginning to take pride in his accomplishments during training and he was anxious to earn the title of U.S. Marine. That was the gist of his explanation to us. We all kidded him about his seeing too many John Wayne movies but our kidding was in good fun and we respected him for his decision.

The twelve of us enjoyed our day of working together and having the freedom of laughing and talking among ourselves. At 1700 hours we arrived back at the barracks to quickly grab our rifles and gear as we headed to chow.

After chow we drilled on the parade ground until 1830 when we ran the dreaded Confidence Course. We were beginning to be able to cruise through the Confidence and

Obstacle Courses with relative ease by now and on most days we ran through both of them at least twice. No longer did any of the recruits in our platoon have a fear of the events that had heights to contend with. Those few recruits that had been unable to conquer their fears were long gone, either to a later arriving platoon or by way of a General Discharge.

After our day's work on the maintenance details most of us were scheduled to stand guard in our Battalion area that same night. I drew the 1200 to 0400 watch. I was awakened by the fire watch at 2300 and fell out into formation on our company street with my rifle for inspection by one of our Drill Instructors who was acting as the Sergeant of the Guard.

After inspection I was marched to my post near one of the barracks in 3rd Battalion. I had a rifle and a twenty round magazine but I had no ammo for the magazine. I had no bayonet either. I marched around and around that barracks building with much trepidation that night. Luckily there was no enemy or intruders to defend against during my watch.

One of my platoon members wasn't so lucky though. While walking guard along a company street an obviously inebriated Corporal approached his post. When challenged to approach and identify himself he shouted out, "My name is George Washington and I'm drunk as hell!" That answer didn't go over with the already tired, frustrated, and angry recruit who proceeded to give him a vertical butt stroke with the butt of his rifle.

After an investigation the recruit was exonerated and after a few hours in the Dispensary the Corporal was released for duty. Now I understood why maybe no recruit walking guard was issued ammo that night.

The next afternoon we were marched to a building where our boot camp recruit photographs were taken. Most of us had been detailed to various late night work details and our pictures clearly reflect the stress we were under at the time.

There were over one hundred of us in platoon 381 and we were shuffled in and out in a hurry as two photographs each were taken, one smiling and one unsmiling.

Private Eddie G. Nickels -April 1966 - USMC Photo

After evening chow a detail of twelve recruits, including me, were taken in a USMC pickup truck to the Weapons Battalion Area which was located about a mile away near the Rifle Range. It was after dark when we were taken there for guard duty so I didn't get a good look at the famous range where so many Marines of lore had trained as Marine Riflemen. In less than a week my platoon would be beginning their training in marksmanship. The trip we made that evening turned out to be interesting for us twelve recruits.

The NCOs in the Weapons Battalion area weren't all Drill Instructors. Some were Marksmanship Instructors on the Rifle Range while others were military police, supply, and communications personnel. It was the first interaction with Marines we had as recruits that weren't engaged in screaming at us. (Except for the Corporal on our prior work detail, who also treated us very decently.) We spent the time when not on guard asking them questions about the remainder of boot camp training. They answered our questions without hesitation and were helpful in relieving our fears of what was to come.

After walking guard around one of the Weapons Battalion barracks I was relieved at 0400. We twelve recruits were once again driven back to our barracks where we had just enough time to shower and change utilities before our platoon fell out for chow at 0515.

After chow we had an hour of PT, then an hour of drill with an emphasis on the manual of arms. We then ran three miles around our running track in boots and with rifles. The three mile run with full packs, gear, bayonets, and rifle was a required physical test that would have to be passed before graduating boot camp. One never got used to the PRT but it had to be practiced every day to get in shape for it.

There was no getting around the Physical Readiness Test. Every Marine had to pass the test before graduating boot camp and every three months thereafter up to the age of forty five. The idea was to make the run so many times that it would seem routine when having to run it as part of the graduation requirement.

The PRT required a three mile run with full combat equipment, including combat boots, rifle, bayonet, full canteen, and fully loaded field pack. You had to finish the run in less than twenty minutes to pass. Another part of the test involved a concrete platform twenty four inches high and twenty four inches wide. You were given three minutes to do sixty two step- ups onto the platform and back to the deck as you counted off. While performing these step-ups you carried the same equipment as with the three mile run. There was a requirement of a twenty foot high rope climb to a head log and back down in sixty seconds, also with full equipment. The PRT event requiring a fireman's carry of a fully combat loaded Marine for a distance of fifty yards was more difficult than one could imagine. The last event in the PRT was an eight foot jump over a deep ditch while carrying a full combat load.

One noticeable change for the better in the attitude of the Drill Instructors was during chow time. We were surprised when on Monday of the first day of the Second Phase of training that the chore of shouting, "Ready, Seats!" was assigned to the last recruit around the table by the Drill Instructor. It was a small bit of change but it signified a slight improvement of the stress level at chow time.

The Drill Instructors no longer stood by our tables and watched us as we ate chow as they did our first five weeks on the Island. I was also grateful that there was no screaming and no nose to nose shouting when a minor infraction was incurred. We were beginning to realize that these small relaxations in stress were granted as we gained seniority in training and as we began to show giant steps in training efficiency.

There was still plenty of physical training during the day and at night on the quarterdeck when we screwed up. Although we didn't see much of the Drill Instructors during the day during our maintenance week we were blessed with their company when we were in the barracks. Each night before lights out some minor infraction would bring down the wrath of the Duty Drill Instructor on all of us.

On Thursday evening at 2145 hours the Drill Instructor held a rifle and equipment inspection just before lights out. After he had inspected the first two or three recruits and their rifles he blew a gasket and shouted for all of us to "Hit the m..... f...ing deck!" Hit it we did and the pushups were many and the side straddle hops were fast and furious. Lights out was scheduled at 2200 hours but there was no rack time for us Maggots/ Shit Birds at the normal time.

The Drill Instructor left the center of the deck where he was watching us sweat out our "bad attitudes" and switched off the lights at 2200. He then said in a lowered voice," Keep your pie holes quiet and continue doing pushups until you make sufficient contrition for your lack of proper respect and appreciation for all the slack you've enjoyed while you were out of my sight this week! You have an obligation to keep your rifles in inspection order at all times! Keep pushin'em up you little p...iesYou'll show some respect for Marine Corps equipment or I'll spend the rest of the night sweatin' it out of you!"

My rack was just a few feet from the hallway leading to the back entry and exit and I noticed someone hurrying down the hallway towards us. As the stranger got closer a recruit near the Drill Instructor's shack screamed out, "Standby, Attention!" With those words every recruit jumped to attention and so did the Drill Instructor.

The strange Marine walked to the light switch and switched it on. As he did so we all recognized Lieutenant Record, the Series Commanding Officer. He walked up to the Drill Instructor and with a frown on his face he harshly said, "What's the meaning of having PT after lights out Sergeant?" The Sergeant replied, "No excuse Sir, we're just catching up on lost time this week." The Lieutenant glared at him and snorted,"It's time these recruits get some sleep. You know this is against regulations. Get them in their racks and get those lights out, now!" With that the Lt. did an about face and walked back down the hallway to the exit.

Without missing a beat the Sergeant muttered in a much subdued voice," Hit the rack Girls!"He then turned on his heels, entered his office and slammed the hatch after him. I think we all feared some form of retaliation from that incident but it never materialized. As usual, we recruits came to attention as we lay in our racks and recited the Rifleman's Creed together before falling asleep.

On Saturday morning April 23, the lights came on as usual at 0330 and some of us were already standing beside our racks fully dressed for the day. As a platoon we were still only allowed two minutes to get into our utilities and have our boots on and laced up. Those recruits that still had an article of clothing remaining to put on after the allotted time had to do a series of PT before they were given another minute to don their clothing or combat boots. The recruits that were already dressed weren't required to exercise. I had, after a couple of weeks, realized that I could save myself some extra PT by getting up five minutes or so before reveille and be standing tall when the Drill Instructor switched the lights on in the mornings.

The first morning I got up early and was standing in front of my rack as the lights were turned on the Drill Instructor looked at me strangely but didn't comment as he stood in front of his desk which was only ten feet from my rack. The next morning several recruits joined me in being ready when the lights were switched on. Afterwards up to a third of the platoon chose to be up and ready early so as to avoid PT. This was a good system for avoiding PT at 0330 in the mornings. Normally Drill Instructors appreciated it when a recruit showed any initiative during training. To our chagrin they eventually tired of having some of us missing out on the early PT and informed us as we hit the rack one night that any recruit standing beside their rack and dressed before normal reveille would have double PT in the morning! I imagine all good things have to come to an end sometime.

One particular Saturday, on April 23rd we hadn't yet been forbidden to be up and ready before reveille. I was out of the rack and dressed by 0300 because of my neck. I hadn't slept much the night before because of a nasty sunburn on my neck. The sun had started being a problem for me and other members of the platoon about two weeks before when the weather started heating up. We had begun tanning some but

I was so light and pale skinned that the sun still did a job on my skin from time to time. I felt miserable but I sure wasn't going to sick bay for a sunburned neck. I would likely have been laughed out of the dispensary for so minor a problem. Like other, cuts, scratches and bruises I just ignored the burning and peeling of the skin. I was in misery for a few days with blistering and peeling of my hide.

The Drill Instructors would normally make the recruits go to sick bay if they detected an illness or medical condition of any kind that a recruit had. They were very conscious of the fact that they were responsible for each recruit's well being and they took this responsibility seriously. On the other hand most recruits tried to hide their illnesses and accidents from the Drill Instructors. They wanted to take no chance of being dropped or set back to another platoon for training. Their main goal was to get their training over with.

I once saw a recruit fall from near the top of a twenty foot rope climb. The Drill Instructors quickly administered to him and were telling him he had to go to the Dispensary to be checked out. He was protesting and saying he would be fine as soon as he "caught his breath." He was still protesting as he was carried away to be checked out. He came back to the platoon the next day and graduated with our platoon. This shows the lengths recruits would go to avoid any kind of medical treatment if it could be avoided. Compared to the fall of my fellow platoon mate my sunburn was nothing.

On Sunday April 24, I got a welcome respite from our normal routine of doing laundry and cleaning chores after church. I was informed after chow that I had visitors from home waiting for me at the Visitor's Center on the main part of the base! Wanda had written me a letter a week before with the information that she and the rest of my family were planning a trip to visit me in a few weeks so I wasn't entirely surprised by the news.

Wanda and I had been married over a year and a half when I was inducted but it still seemed as if we were newlyweds and I had really missed her the five weeks I had been gone from home.

Right at 1200 hours the Drill Instructor gave me permission to leave the barracks and walk to the parade deck where my family was waiting. They could hardly believe the change in my appearance in just five weeks when I walked up to the car. I was so tanned that I looked like I had been lying on a beach the past five weeks instead of on a drill field at Parris Island. I walked with my shoulders back and my head looking straight ahead as I was used to doing while marching as the Drill Instructors called cadence." A few weeks on Parris Island would change anyone," I commented when Dad mentioned how much I had changed.

My Mom broke out in tears when she saw me and Wanda had a big grin on her face. The first place Dad wanted to take us when I got into the car was to the PX to get me something to eat and drink but I explained how we still weren't allowed any extras during training and would suffer dire consequences if caught doing so. He said to me,"I understand about rules and regulations and you're right in not wanting to break them. During my time in the Army I felt the same way." Mom just said, "Oh Ed! One bottle of pop won't cause him any harm, surely!" She then asked me if I was sure I didn't want something to eat or drink and I assured her that I was fine. They had already planned to have a picnic when they came to see me and had some food they had brought with them for the purpose.

After some discussion Dad started the car and we proceeded to the causeway leading off the Island. We traveled back along the causeway in the opposite direction the bus had traveled when we first reached Parris Island several weeks ago. Just before we came to the bridge which spanned Archer's creek we came to a picnic area on the right side(starboard side as I knew it then) of the road. Dad had noticed it as they drove along the causeway and thought it would be perfect for the picnic.

My brothers and sisters also come along on the trip and I enjoyed their questions about our training at Parris Island. All of them were young then and they wanted to know how rough the training was, did I have a rifle yet, and how many men were in my platoon? There were other questions but I had as many questions about home as they did about the Island.

Wanda and I walked and talked among the picnic tables while the rest of the family was eating. We discussed our financial situation and how I was going to try to get a place for us wherever I was posted after my six months scheduled training was over. Other than that basic statement I couldn't offer her any assurance of our future other than I would send her some money when I got paid. She never once complained about her situation or asked about what she would do without funds. She was, and is, always willing to make do on what we have.

Visiting hours were scheduled from 1200 to 1600 hours so we started back towards the 3rd Battalion parade deck at 1530 hours. The goodbyes were sad but necessary and Dad promised to bring the family back down for graduation in a few weeks. After hugs all around I started back towards the barracks area with a big lump in my throat. As I walked away I watched as Dad pulled away with Wanda, Jimmy, Rita, and Phil waving from the back window and Mom and Kathy waving from the front seat. My sister Marlene was married and living in Michigan or I'm sure she would have been along also. The support of one's family is priceless when we are separated by distance or time. I know it meant the world to me at the time.

Chapter Twelve

GRASS WEEK

Move to Rifle Range

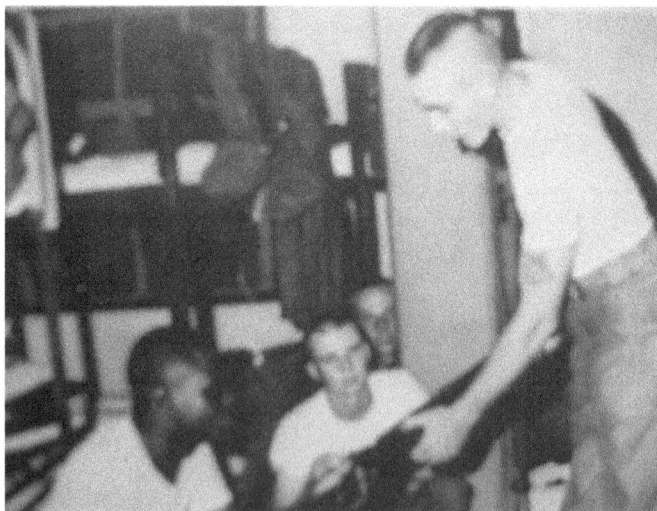

From left: Pvt. Walton, Pvt. Doty, Pvt. Kinder, Drill Instructor Sgt. Shue, Platoon 381 Squad bay. April, 1966 (USMC Photo)

After the visit of Wanda and my other family members I felt a little depressed instead having my spirits lifted that evening. I was very glad that they were able to visit but the sadness of seeing them have to leave was almost overwhelming to me.

I didn't have long to mope around after returning to the barracks as we marched to chow just a few minutes after my return. After chow we drilled and practiced the manual of arms for an hour or so before heading back to the barracks.

As we stood at attention in front of our racks the Drill Instructor informed us that we would spend the rest of the evening engaged in a field day. This was in preparation for our scheduled move to the Weapons Battalion area where we would spend the next two weeks on the Rifle Range.

We normally dreaded a field day of scrubbing and cleaning the squad bay from top to bottom but knowing we were headed for the most important aspect of our training as Marines gave us a needed shot of adrenalin to heave to with a will as we worked. We were finally going to be able to fire our rifles after more than four weeks of familiarization and training with them.

As we marched and trained in the 3rd Battalion area we could easily hear the sound of heavy firing from the Rifle Range each day. The sound of hundreds of rifles firing at the same time sounded exactly as I had remembered reading about in Civil War books and magazines over the years. The sound was similar to the tearing of paper, in my opinion. Many witnesses to Civil War battles described the sound of so many rifles firing at one time as of paper tearing. I was surprised that they had the description almost exactly right.

When the cleaning was completed at 2100 we got our seabags out of our lockers and emptied our foot lockers of everything we would need on the Rifle Range. Towels, wash cloths, extra skivvies, two set of utilities, toilet articles, and just about everything we owned was packed into our seabags and our field packs for the move.

Monday morning reveille was at 0315 and we were at the chow hall by 0415. I had begun to like the food very well and had even had acquired a taste for the powered milk that was dispensed from five gallon paper cartons inside the dispensers. I had eaten at some of the other mess halls before our platoon assembled at our 3rd Battalion barracks and I'm convinced that we had the finest mess hall on the Island. Even today when I tell other former Marines the battalion I trained with, they'll nearly always interject with, "Oh, so you trained in the Hollywood section of Parris Island?"

Our march to the Weapons Battalion area was going to be in conjunction with a conditioning hike, so we had a fully loaded field pack, helmets, bayonet, full canteen and cartridge belts with two M14 magazines in addition to our heavy seabags to hoist onto our backs that morning. We started the march at 0515 hours going west on Wake Boulevard but we quickly filed to the fields on the port side of the road and traveled through the trees and fields on a well worn dirt path. After a few minutes of hurried marching the humidity even at this early hour began to be bothersome and annoying. I remember thinking that I was glad we weren't marching with our heavy loads in the hottest part of the day.

About one and one half hours after leaving our 3rd Battalion area we came to a row of brick barracks on the port side of the road and five -half mile deep rifle ranges on the starboard side of the road. These five ranges are known as Able, Baker, Charley, Dog, and Echo. Our series of three platoons would be assigned one of these ranges to do our firing on when we started our firing practice next week. This week we would be dry firing or "snapping in" all week to become familiarized with the rudiments of marksmanship before we advanced to the actual firing of our weapons.

Between the rifle ranges and the Wake Boulevard were several small pavilions where the Marksmanship Instructors would give us classes each day between periods of dry firing. Between the pavilions and Wake Boulevard were the grassy areas where we would spend most of the next six days aiming and pretending to shoot at the target barrels. Each barrel has small black silhouettes painted on it that match the silhouettes on the Rifle Range targets. We would spend many hours the coming week in the prone, sitting, kneeling, and offhand (standing) position while aiming our M14s at these barrels.

After arriving at our assigned barracks we hustled up to the second deck and deposited our gear on one of the squad bay racks. We then ran back to the street with our rifles, bayonets, and cartridge belts and marched across Wake Boulevard to one of the pavilions to our first class on marksmanship.

All four of our Drill Instructors had accompanied the platoon on our conditioning march from the 3rd Battalion area to the Rifle Range and would rotate shifts so that one of them would be with or nearby the platoon at all times. While attending marksmanship classes or during actual firing on the range we would be charge of our Primary Marksmanship Instructors (PMIs)who would be responsible for our safety and discipline.

They applied this discipline very liberally on occasion and had no qualms about delivering a swift kick to the butt or a sharp rap on the helmet when a safety rule was violated. This was against regulations of course but when it came to safety, regulations came second, in my humble opinion.

While lying in the grass aiming at a white barrel for eight or ten hours a day with a rifle sling tightly wrapped around your upper arm so as to nearly have the circulation cut off, it can be tempting to doze off or at least relax for a moment. This would likely bring a swift and unexpected kick to the rump that had a tendency to wake you up, hurt your feelings, and hurt like the devil. There were very few recruits that escaped a kick or two while on the Rifle Range over the two week period. Sometimes these kicks were undeserved but when a rifle barrel was pointed in a direction other than where we had been directed to point them, the PMIs wanted to get our attention immediately to make sure it didn't happen again.

If a safety violation is serious enough a recruit can be ordered off the range for the rest of the day and face the Drill Instructor's wrath or maybe even dropped to a platoon that will be starting their Grass Week at a later date. The violation that will more than likely trigger the latter punishment was when a recruit pointed a loaded rifle at another human being while on the range. Even pointing an unloaded rifle can bring the same punishment. Safety is paramount and is taken very, very seriously.

When we entered one of the pavilions for our first class to be given by a PMI that Monday morning we received our first instruction in range safety and shooting procedures. The PMI explained that we would have to learn to fire our weapons while our bodies were in a contorted position for most of the shooting positions. The sitting position of firing is the one that presents the most painful and difficult position to achieve. First you have to cross your ankles while standing and lower yourself to the ground with your body at a forty five degree angle from the firing line to the target. You then must lean your body forward and rest your arms on the shin bones. With your left arm vertical and your right arm parallel to your right shoulder you pull the butt of the stock against your cheek while your stomach begins to try to cramp up on you. This causes one to impulsively straighten out the legs which means you'll have to once again suffer as you get back into position.

This is why a whole week is devoted to practice in training your muscles and body to adjust to getting into position for firing for qualification. It would be almost impossible to just show up on a Marine Corps Rifle Range and try to achieve a passing score in marksmanship. As difficult as the week of snapping in turned out to be, the members of my platoon were grateful for the many hours of instruction we received from the PMIs when we were on the actual firing line the next week.

The four shooting positions used in qualifying on the Rifle Range were;

- The Prone or lying down position- The sling is looped and tightened high on the left arm. Dropping to your knees you place the butt of the rifle on the ground under the center of the body and pivot down to the left side of the rifle. You then place your left elbow directly under the rifle. Pull the butt of the rifle into the shoulder. Keep your feet apart and keep your shoulders level with the ground. Place the magazine in the rifle and release the bolt latch. You're then ready to fire the rifle.

- Sitting position-The sling is looped high on the arm and dropping to the ground you place the upper left arm inside the left knee. The butt of the rifle is placed against the right shoulder and the lower right arm rests inside the right knee. Place the magazine into the rifle and allow the bolt to go forward.

- Kneeling position-Drop to the right knee. Place the right leg parallel to your body with your left foot towards the target. Lower your buttock to the right foot and rest the left forearm on the left knee. Put the rifle stock against the right shoulder. Insert the magazine into the rifle and allow the bolt to go forward.

- Standing position-Spread the feet to a comfortable position. Place your left hand in a comfortable support position on the rifle. Place the rifle into the right shoulder. Hold the right elbow high to form a pocket for the butt of the rifle

The sitting position is by far the most difficult position to assume and the stress and strain it requires of your muscles and tendons can be mastered only by repetition. The kneeling and standing positions are slightly easier to assume but accuracy suffers because of the stability in holding the rifle. This was not at all the way I had learned to shoot with my .22 rifle when I started hunting with it at age twelve! I almost had to start from scratch just like most of the city boys in our platoon that had never fired a rifle before.

Our first stint of dry firing while aiming at the small barrels started the afternoon of our arrival at the range. We had been taught in our second class of the day how to tighten our rifle slings on the upper part of the left arm when getting into one of the firing positions. One thing that couldn't be taught however, was how badly the pain and lack of circulation in your arm would affect you. The stretching of muscles while getting into position was another adventure in patience and endurance. It made us all miserable that first day and the heat, mosquitoes, and sand fleas were other unbearable factors to consider.

There were no water fountains on the range. We had the tap water in our canteens which we filled from the sinks in the head when we deposited our gear in our barracks across the road. We had drained our canteens by the time we marched back across the road to noon chow. We again filled them up at the barracks when making head calls after noon chow.

It didn't take very long for that water to become hot in the canteen also. Some recruits were lucky enough to have been issued plastic canteens but I had an old beat up aluminum one that appeared to be surplus from World War Two. The water was hot but wet and we drank it anyway. By early afternoon several recruits were out of water and the sun was beating down mercilessly.

We were scattered in a circle around the barrels as we aimed and pretended to fire at the target. I'll never forget how humid and hot it was that first day. We labored there in pain, misery, and heat as we got into the four firing positions time after time. We would spend a half hour in the prone position or sitting position then switch to a different position for another half hour. This went on all afternoon in the sweltering sun.

If any of the platoon members had an idea that we were in for an easy week ahead of us they were quickly disavowed of those thoughts by the end of the first day. We were sore, bruised and slightly sun burned as we filed into the mess hall for chow at 1730. At least the chow was almost as good as we had enjoyed at the 3rd Battalion mess hall. I probably came close to drinking a full gallon of that fine powdered milk that evening. At least it was cool and wet.

The Rifle Range activities didn't excuse us from our regular training duties. After chow we took our rifles outside the barracks and cleaned them in the large wooden laundry racks that were located there. Afterwards we had drill and PT before heading to the second deck of the barracks to begin our nightly cleaning details. We also had to fold and stow our gear in the footlockers and wall lockers. By 2100 we had our gear stowed and the barracks scrubbed up nicely.

Mail call was held at 2130 and I surprisingly received a letter from my former employer, J Don Collins. It was nice letter that served to raise my morale, like all of the letters I received tended to do. Don had a habit of giving nearly everyone he met a nickname and he had given me the moniker of "Mayor" of my residential area, (Tunnel Hill.) Because I was now in the military he began the letter with "Dear General." I had received quite a promotion since my arrival on the Island! I was always called "General" by him afterwards.

J Don had served in the Navy during World War One. He and I spent many hours discussing the military while we were working together in his super market. Don's son- in –law, Bert Francis, was a Captain in the Marine Corps and I had met him just before I left to be inducted. I had no idea at the time that he and I would soon be fellow Marines or I would have surely discussed my situation with him.

The next morning it started all over again. We had chow at 0430, then had drill and PT while it was still dark outside. We then ran a three mile conditioning run before heading back to the pavilions to begin more classes. It felt good just to sit down for a spell while the Instructor was engaged in teaching us some fundamentals of marksmanship.

After a couple of classes we once headed a few yards away to the by now familiar barrels and assumed our first painful dry firing position of the day. It didn't take long for our muscles and tendons to begin their protest of our contorted positions .The hot sun sure didn't help matters any either. Soon one or two recruits jumped to their feet and started trying to rub out the cramps in their arms, stomach, or legs. Naturally that brought out the ire of the Drill Instructor observing nearby. Standing up or any other deviation from orders to concentrate on snapping in was strictly forbidden.

"Get on your feet, get on your feet and assume the position of attention," he barked. We appreciated the opportunity to stand but we didn't care for the exercise which followed. We were ordered to loosen the tight slings from around our upper arms that were almost cutting the blood circulation completely off. Then we went to "port arms" and were instructed to raise our eleven pound rifles straight up to an even level with our shoulders with both arms straightened while holding our rifles.

This wasn't our first exercise of holding our rifles in front of our bodies until the biceps cried out for relief. We had done this many times in a barracks setting for some infraction or other. However, this particular exercise that day was memorable due to the fact that our muscles were already sore and ailing from the previous day's snapping in activities. It seemed that we held our rifles for at least fifteen or twenty minutes with every neck muscle we had straining to compensate for our sore arms.

After a sufficient length of time had elapsed we were instructed to assume our prone position on the grass and adjust our slings back on our upper arm. After that experience no one jumped to their feet with a cramp. If a cramp hit we just grimaced and groaned under our breath.

Around 1700 we went on a conditioning run with rifle, full canteen, helmet, and helmet liner. Lying in the grass the past two days and making our bodies conform to almost impossible firing positions had our muscles aching almost beyond belief. The Drill Instructors had in mind just the exercise needed to "unkink" our muscles.

We ran with our rifles at port arms for at least five miles as we circled the Weapons Battalion area. We circled some areas several times as we ran in formation while the Drill Instructor shouted cadence. He would call the cadence out and we would repeat the words while shouting as though our lives depended on it .A couple of the lines in those cadences went like this;

 " If I die in a combat zone, box me up and send me home!"

 " Mama, Mama, don't you cry, Marine Corps Motto is Do or Die!"

Singing a witty cadence is good for morale and helps keep the whole platoon of over a hundred men in step during a tough conditioning run. There's nothing more motivating than a hundred voices singing a ditty while marching or running in step down a dusty road in the middle of the boondocks while the sweat is pouring into your eyes.

At the beginning of this fifth week we had finally been extended the privilege of unbuttoning the top button of our utility blouses. Unless one has had the experience of marching and running all day with the top button of a tight jacket buttoned, it is difficult to know the way we felt when the Drill Instructor allowed us that simple relief. Not only did we feel more comfortable but we were now marked as Phase Two Recruits to incoming new recruits.

There was a noticeable ease of extreme punishment and stress during our "Grass Week" also. The Drill Instructors and Marksmanship Instructors wanted our whole attention while on the rifle range. "Quarterdecking" and extra PT was kept at a minimum and stress levels took a dive as the morale level correspondingly increased during the week.

Drill Instructor Sgt. Garcia's favorite tactic to punish minor infractions was to use four fingers of one hand to thrust into our stomach up to the first knuckles of his fingers. I had been the recipient several times over the past weeks of his fingers making my stomach touch my backbone when I had committed a minor infraction of some sort. The absence of this mode of punishment allowed my sore stomach to rest a spell during Grass Week.

Chapter Thirteen

FAMILIARIZATION AND TRIANGULIZATION

The photo on the opposite page is one taken of me on the Parris Island rifle range in April;1966.This is on the 500 yard line with the targets barely visible in the distance. I have on a padded shooting jacket, or vest, shooting gloves, and of course my trusty M14 rifle. I am aiming my rifle from the "offhand" or standing position. Notice that one end of the rifle sling is wrapped around my left arm. (USMC PHOTO)

Just because we were on the rifle range didn't excuse us from our regular Thursday field day duties. At 2000 hours the Drill Instructor shouted," Platoon, Attention! "Port side will field day the squad bay and the starboard side will field day the head and my house (office). Squad leaders assign your squads to their details. I better not hear any talking, bitching, or any other discussion while working. When I give the word all I wanta see are a.. h...s and elbows on the deck and hard at work! Do you understand me?" We all screamed together, " Yes Sir!" "Then get to it!" he shouted. More than one hundred voices than rang out with a loud "AYE AYE SIR!"

The squad leaders assigned us our job detail and with the Drill Instructor observing we scurried about in a run to retrieve our cleaning gear. I was assigned along with three other recruits to sweep the squad bay decks and to dust the racks afterwards. I always enjoyed these Thursday field days as it gave us recruits an opportunity to move around the barracks freely during the one and one half hours allocated for the cleaning. Anything beat having to stand in front of our racks at attention while the Drill Instructors harassed us about a trivial miscue that we had committed during the days training activities.

If we weren't standing at attention we were engaged in PT in front of our racks or on our knees slapping our palms against the deck to toughen them up. A little cleaning detail was a pie job in comparison.

At least one Drill Instructor was always on duty in the barracks with us. He never left our sight unless he had a few minutes of paperwork to catch up on or maybe was busy correcting a recruit in his office. When a recruit was called into "my house" as the Drill Instructors called his office, shouting, screaming, and sometimes what sounded like scuffling was sure to follow. There was never a dull moment on Parris Island while in boot camp.

As we began our cleaning, the Drill Instructor went into his office and closed the hatch behind him. We took advantage to begin whispering to our fellow recruits and catching up on the latest scuttlebutt. All of us were whispering our heads off trying to socialize while we had a chance. All at once we were interrupted by a loud scream, "Aiiiieeeeeeeeeeeeeee!"

With my broom in hand I glanced up to see a recruit running down the squad bay with a bayonet in his hand while shouting, "Leave me alone, I'm going to end all this bullshit! I'm going to end it!"Chasing him were two of the squad leaders and several other recruits who shouted for us to "Get out of the way, he's going to hurt himself or somebody else!"

He stopped running directly in front of the three or four of us that were sweeping under the racks and as I looked into his eyes he looked as wild as I had I had ever seen anyone look. I understood instantly that this recruit had just lost his senses under the stress and was easily capable of harming himself or someone else while in this state. He was crying with foam running down the sides of his mouth as he was lashing out left and right with his flashing bayonet just five feet away from where we were standing.

Before anyone else could react, those recruits that had been chasing him caught up with him. By coming in behind him a couple of them grabbed his hands and others wrestled him to the deck and took his bayonet from him. He was bawling and babbling nonsense as they hustled him away and into the Head so he could be calmed down out of the sight of the Drill Instructor.

I could never figure why the Drill Instructor never heard anything. Maybe he did, and just let the recruits handle the situation as a training tool. The stress of boot camp was unlike any that the recruits had ever faced and some reached their breaking point sooner than others. This particular recruit had never before given anyone any trouble but he had apparently reached his limit. The threat to harm himself and anyone else that tried to stop him was, of course a cry for help. After this incident he continued training without any other incidents. No one informed a Drill Instructor of what had occurred.

The recruit was one of our platoon's pickup recruits that had been set back from another platoon. He was only seventeen years old at the time of the incident. He was not the first one of our platoon to threaten suicide while training. There had been several others before him. Most of them had been discharged. Some recruits had even gone to the extreme of telling the Drill Instructors that they were homosexuals so they could get out of training and the military. This method would usually work as no homosexuals were allowed in the military during those days. Sometimes a recruit would declare he was gay and would be taken away to be evaluated and we would later see that recruit marching with a newer platoon. Evidently their declaration had been a ploy to get out and they had been found out.

Recruits often would try to swim off the Island to the mainland when they had experienced the trauma of boot camp for a few days, especially the first two weeks or so. One night when we were in our 3rd Battalion barracks a recruit was brought into the barracks wearing soaking wet utilities and was under guard. He had been apprehended near the front gate after swimming Archer's Creek while trying to escape from the Island. No doubt some of the recruits that had gone A.W.O.L. had managed to escape in this manner. We had been told in our first week that "There were plenty of gators in the waters surrounding Parris Island and if the MPs don't get you the gators will!" I never had the urge to test that theory myself.

One common saying among our Drill Instructors was that, "There are two ways off this island. One way is on a bus and the other way is in a box. The way you leave here is entirely up to you scumbags!" This scumbag personally preferred the bus to a box.

Saturday morning April 30, we jumped out of the rack as usual at 0330 and stood in front of our racks as we sung the Marine Corps Hymn. We had done this every morning of boot camp and we were expected to know the words to the Hymn and to sing it at the top of our voices. It was almost comical to see one hundred or so young men in their skivvies standing on both sides of a squad bay and singing loudly at attention while the Drill Instructor strutted up and down the squad bay making sure everyone was shouting the hymn to the rooftops and not just moving our lips.

It was somewhat embarrassing at times for the recruits to have to stand at attention when first getting up in the mornings. After all, we were typical young men in the prime of life and while singing we were usually involuntarily saluting the Drill Instructor with an appendage other than our arms and hands. Since this is a true memoir of boot camp incidents I have to be honest about incidents that really happened during training. The Drill Instructors ignored the problem and to their credit never gave us a hard time over it.(No pun intended.)

After PT and our morning conditioning run were marched to the 45 Caliber Pistol Range where we were scheduled to take Familiarization training with the pistol. We would at last get to fire our first weapon on Parris Island this day.

Our platoon wasn't the only one taking the course this day. The other two platoons in our series would be with us in the classes and on the firing range. Platoon 380 and 382 were as large as our own 381 which made a class of over three hundred recruits in the bleachers for the pre-firing class given by the Pistol Primary Marksmanship Instructor.

The Pistol Familiarization Training wasn't nearly as extensive as the rifle instruction given to the recruits during boot camp. After all, every Marine is a rifleman but not every Marine would be required to carry a pistol in battle.

The 45 caliber M1911 pistol is a single action, semi-automatic magazine-fed recoil-operated handgun chambered for the .45

ACP cartridge which served as the standard –issue sidearm for the United States armed forces from 1911 to 1985.[1]

The .45 caliber pistol weighed 2.44 lb and the detachable magazine held seven rounds. The pistols we fired that day were all but worn out from so much use on the Familiarization Course and it's a wonder that any of us were able to satisfactorily hit any of the targets. Some of the internal parts of the pistol I fired were so worn that they rattled when the weapon was moved around.

The targets were set up at fifteen and twenty five yards. Each recruit fired around fifty rounds total. The magazines were usually loaded with five rounds each as we fired on each target.

The Marksmanship Instructors were as strict as the Drill Instructors while we were on the firing line. The reason for their strict behavior was the dangerous handling characteristics of the .45 Caliber Pistol. The pistol was large, heavy, and had a kick that was almost like a bucking mule. When fired, the recoil would throw your arms and hands back towards your head. The poor condition of the pistols would cause many of them to have a hair trigger which would sometimes cause a recruit to inadvertently fire a round into the dirt or into their own ear or fingers. The recruits would sometimes grab the barrel of the pistol as they racked the pistol for another shot and would blow off a finger.

---

[1] http://en.wikipedia.org/wiki/M1911_pistol

Because of these incidents the Instructors had little tolerance for mistakes on the range. They were kicking rumps all over the range that day. It wasn't legal but the kicks and punches probably saved a lot of accidents while on the range.

I came very close to getting my own rump kicked while firing the Familiarization Course with the .45 Caliber pistol. I had already fired thirty or forty rounds at the targets which were25 yards away from the firing line when the Marksman Instructor gave the command, "With 10 rounds, slow fire, ready on the firing line! Ready, fire!"

There's almost always one or two recruits that either fail to grasp the message or gets the message wrong in some fashion or other. This day I was one of those recruits. For some unexplained reason I thought I heard the Instructor say, "With 10 rounds, rapid fire, ready on the firing line! Ready, fire!"

I had two magazines with five rounds in each and I quickly fired five rounds at the target, released the magazine catch, and hastily inserted the other magazine. I then fired the other magazine at the target as fast as I could pull the trigger. As I finished firing I lowered the barrel towards the deck and locked the safety. I still heard some firing when I finished so I turned towards the Instructor to see if he had noticed how quickly I had fired all 10 rounds. I was expecting an attaboy from him. I thought to myself, " Man, we country boys can show these city boys a thing or two about handling a firearm if they'll just watch us." It was then that I heard the Instructor shouting, "What in the &@&5$*^% hell are you doing recruit? Don't you believe in following instructions? You better get with the plan, Numb Nuts!" I looked around to see who he could possibly be talking to. I knew it couldn't possibly be me, but wait, why is everyone staring in my direction? I thought it must be someone on the other side of me that he was talking to so I turned around to see who the culprit was that had screwed up. I saw that those beside me were also looking in my direction. It then hit me that I was the object of attention of the whole firing line. As I glanced back at the Instructor he again let me have it. " That's right Sweetheart, you're the one that f...ed up! Don't be looking around trying to blame someone else! Now, you just stand there at attention while the rest of the platoon follows my instructions with 10 rounds of *slow* fire!" If I would have had my druthers I would have rather had the usual kick to the backside instead of the humiliation of not only being the first recruit to screw up on the .45 Caliber Pistol firing line, but my first time to screw up with the whole platoon watching. My feelings were hurt because I thought I did rather well to get 10 rounds off so quickly.

After noon chow our series of platoons marched to the 1000-inch Triangulization Course where we would finally get some familiarization training with our M14 rifles. We had been anxious to fire them since the beginning of Boot Camp and now we would finally get our wish.

We would be on the rifle range next week and the Triangulation Course would help us to get the windage and elevation set on our individual weapons. By understanding how to apply a certain number of "clicks" to the front or rear sights of our rifles we could get our "Kentucky windage" set so as to try to keep our shots "in the black" of the targets.

If your rifle is shooting low on the target the front sight needs to be lowered on the rifle. Each click of the sight moves the point of impact one inch for every one hundred yards. If your shots are high on the bulls eye the front sight needs to be raised one or more clicks. Again, each click moves the impact one inch for every hundred yards.

Conversely, if the rifle is grouping your shots to the left or right of the target bull's eye, the rear peep sight is adjusted one click left or right which changes the point of impact one inch for every one hundred yards.

The wind is always a factor when firing a weapon but because reading the wind is an inexact science, using the correct number of clicks on the rear peep sight is mostly guesswork. There are red flags located on poles at each end of the range when firing is in progress. Wind direction is obvious when observing the flags and wind speed is estimated by the degree to which the flags are whipping in the wind.

These facts were drummed into us during Grass Week by the Marksmanship Instructors in almost all the classes we attended during the week. As basic riflemen all Marines are required to be able to aim at a distant 500 yard target and set his own elevation and windage on the spot. He can then make adjustments as needed after firing a round or two.

When our platoon reached the Triangulization Course Saturday afternoon we were given a quick refresher class on setting our correct windage and elevation when we fired our rifles on the 1000-inch line. We then proceeded to the firing line and finally fired our M14- 7.62mm rifle for the first time. As we fired from the 1000-inch line we each were able to set the windage and elevation on our individual rifles. We were now ready to advance to the 200,300, and 500 yard line targets on the Rifle Range on the following Monday.

The 7.62mm M14 was adopted by the U.S. military in 1957 to replace the 30.06 M1 Garand of World War Two and Korean War fame. The M14 was in service for less than 10 years before being replaced by the smaller caliber 5.56mm M16, beginning in 1966.The Marine Corps continued using the M14 in recruit training until December, 1971.

The M14 is a rotating bolt, gas operated, air cooled, magazine fed shoulder fired weapon. It is 44.14 in. long and weighs 8.7 pounds. With a full magazine and sling the weight is 11.0 pounds. The effective maximum range is 503 yards (460 meters).[2]

The M14 could be fitted with a selector switch on the side that allowed a cyclic rate of fire of 750 rounds per minute. The magazine capacity was 20 rounds of 7.62mm ammunition.

A Marine rifle squad had three fire teams of four men each. One man in each fire team was designated as an automatic rifleman and was equipped with an M14 with a selector switch. Each squad also had a squad leader, which made a full squad consist of thirteen men.

The M14 was, and is, in my opinion the best rifle ever produced,(although the M1 Garand is a close second). I found it to be accurate, well balanced, and packing a powerful punch.

_____

[2] http://www.usmcweapons.com/M14-Rifle-in-the-US+Military

Firing it for the first time caused several bloody noses in our platoon that Saturday afternoon.

There was one particular recruit in the platoon whom was above average in almost every exercise we performed in boot camp but he faltered badly at first on the Rifle Range. The first shot he took on the 1000 inch line bloodied his nose and bruised his right cheek. He was very downcast about the incident and took some good natured ribbing from the other recruits the rest of the day.

Private Betts, a member of our platoon that had been in the hospital with diagnosed mononucleosis for the past two weeks rejoined our platoon in time to participate in firing the Triangulization Course this day. We didn't yet know it but the still recovering recruit would end up firing one of the highest scores in our platoon during the final qualification day. It was a major accomplishment for someone that had missed most of the classes and instruction on marksmanship during Grass Week. The Drill Instructors were so pleased with his effort that they called him out before the whole platoon to give him proper credit for his dedication and determination to succeed under adverse conditions.

Chapter Fourteen

QUALIFICATION WEEK

My Rifle

THIS IS MY RIFLE. There are many like it but this one is mine. My rifle is my best friend. It is my life. I must master it as I master my life.

My rifle, without me is useless. Without my rifle, I am useless. I must fire my rifle true. I must shoot straighter than my enemy who is trying to kill me. I must shoot him before he shoots me. I will . . .

My rifle and myself know that what counts in this war is not the rounds we fire, the noise of our burst, nor the smoke we make. We know that it is the hits that count. We will hit . . .

My rifle is human, even as I, because it is my life. Thus, I will learn it as a brother. I will learn its weakness, its strength, its parts, its accessories, its sights, and its barrel. I will keep my rifle clean and ready, even as I am clean and ready. We will become part of each other.

We will . . .

Before God I swear this creed. My rifle and myself are the defenders of my country. We are the masters of our enemy. We are the saviors of my life.

So be it, until victory is America's and there is no enemy, but Peace!

OPPOSITE: This photo is of a Marine Recruit aiming his M14 rifle while in the offhand position on the Rifle Range at Parris Island. I'm not sure if this is a picture of me or another platoon member, although it appears to be me. Note the right arm held parallel to the rifle and the rifle sling is tightened around the left upper arm. This is the correct stance while firing in the offhand position. The photo is from my recruit graduation book. (USMC PHOTO 1966)

On Monday morning May 2, 1966 reveille was held at 0300. We jumped from our racks as soon as the lights were switched on by the Drill Instructor. The yelling and screaming of the Drill Instructor was no longer necessary to shock us awake at so early an hour, as we had trained our minds and bodies to awaken the instant the lights came on. [1]

We jumped to the deck and snapped to attention, still in our skivvies, and sung the Marine Corps Hymn with even more gusto than usual this morning. After all, this week would be one of the most important in our careers. To be a fully qualified rifleman was essential in a Marine's life and career. To fail to qualify on the Rifle Range was the biggest failure a Marine Recruit could suffer during boot camp. Failure would mean being set back two weeks or so in training and having to suffer through Grass Week and Qualification Week all over again. To fail a second time could result in being dropped from training

---

[1] I'm still in the habit of waking up at the sound of a light switch being switched on or when a light comes on in my room although it has been 47 years since Boot Camp. I guess I still fear a rap on the noggin if I don't!

altogether. No one wanted to be a failure, not even members of a platoon such as ours that were almost all draftees. We were proud of our surviving the rigors of Parris Island thus far and we wanted to succeed as much as any volunteer we were training with.

This would be our first day of actual firing at the targets from 200, 300, and 500 yards. The 1000 inch Triangulation Course we had fired this past Saturday was now behind us. Today and the rest of the week we would get to fire the Rifle Range course each day. Friday would be the day that counted and that would forever be noted in our individual record books.

The Rife Range Course was known as the "Known Distance" (KD) Course. The distance we would be firing from was 200, 300, and 500 yards. The targets were different for each yard line and were known as Able, Baker, and Dog targets. The Able target was at the 200 and 300 yard line and would be the first one fired each day. This target was six feet high and four feet wide. It had a twelve inch bull's-eye in the center of the target.

The next target was known as the Dog target and was six feet square. It had the low profile silhouette of a head and upper body in the center of the target. The silhouette was 19" tall and 26" wide.

The target at the 500 yard line was the Baker target and it had a larger profile silhouette of a head and torso. The silhouette was 40" tall and 20" wide.

The course had a 250 possible score if every shot was a bull's-eye. A total of fifty rounds of 7.62mm ball ammunition were fired at the three targets with a bull's-eye being worth five points and four, three, or two scoring points if a shot hit in a lower corresponding ring of the target.

The fifty rounds were fired as follows:

Able target- 200 yards-5 rounds sitting,5 kneeling, 5 standing

Dog target- 200 yards- 10 rounds rapid firing sitting

Able target- 300 yards-5 rounds sitting

Dog target-300 yards- 10 rounds rapid fire prone

Baker target-500 yards 10 rounds prone

The recruits working in the butts (where they would pull the targets down and mark them after each shot) would exchange places with their fellow platoon members when they finished firing the course. One half of the platoon fires the course while the other half work in the butts, pulling down and marking the targets. The butts are located near Ribbon Creek where six Marine recruits were drowned while on a night punishment march on April 8, 1956.Spent rounds from the rifle range would fall harmlessly into the swampy area near the creek.

The butts are where the fifty targets for each firing range are located. They consist of a large concrete pit with each set of targets mounted on wooden boards. When a round is fired by a recruit the recruits assigned to the butts will pull the target down and mark the impact area with a 6" white or black spotter disk on the newly raised target.

Sometimes a target was completely missed and the Range Officer would announce loudly for the pit crew to lower and check the target for a hit. "Mark number two, or mark number twenty!" was a common refrain from the Range Officer to the waiting Pit Crew. If no hit was discovered, a long pole with a red flag tied to the end was waved from left to right across the target signifying a missed shot. This flag was known as Maggie's Drawers. No Marine wanted a Maggie's Drawers if at all possible.

Working in the pits could be very dangerous to one's health. Often an errant shot from the firing line would hit a wooden arm of the target side or graze the top of the concrete barrier overhead. The "snap" of the rounds hitting the targets mix with the loudspeaker of the Range Officer and the shouts and catcalls of the recruits in the pits, making for a place of confusion and some definite fear while working the pits.

Accidents in the pits were few and deaths were even fewer. There had been only one fatal accident up to that time to my knowledge and that incident happened to a recruit "pulling butts" at Parris Island during the Korean War.[2]

As our platoon approached the Range that morning I couldn't help but be nervous about firing the course for the first time. We began on the 200 yard line at 0630 as the first and second squads were assigned to the fifty or so numbered firing positions on the 200 yard line. The third and fourth squad were marched to the butts to labor there while their fellow platoon mates fired the course first. I was in the third squad so I spent

---

[2] Da Cruz, Daniel. *Boot*, 148

the first half of the day pulling butts and had to wait until the afternoon for my turn at the targets.

Our two squads stacked our rifles and filed Into the concrete pit where our Drill Instructors would stay with us as we worked. After some instruction on how to pull and mark the targets we waited a few minutes before the firing started. At first it was difficult not to duck at the sound of a round striking the concrete barrier overhead but we soon got used to the sound and went to work with a will.

The mosquitoes made a feast of us as we worked and the heat was stifling in the nearly enclosed area of the pits. The only relief we had was when the firing relay moved from the 200 yard line to the 300 yard line and as they moved later to the 500 yard line. We were then allowed to relax for a few minutes.

It was after 1200 (noon) before the first two squads finished firing on the course. Chow was brought to the field to us and consisted of bologna and salami sandwiches, an apple, banana, or orange, and a pint of milk. It wasn't exactly my favorite choice of foods and I gave my salami sandwich to another recruit. I ate a banana and drank the milk which wasn't the best combination of food groups.

After chow our two squads switched places with the first and second squads and proceeded to the 200 yard line for our turn at the targets. There are fifty numbered firing points and each of us was assigned a numbered firing point which corresponded with the number of the target we would be firing on. All the targets looked alike and it was fairly easy to fire on the wrong target at times if one didn't pay close attention to what they were doing. Firing at the wrong target would result in lower scores and maybe failure of qualifying altogether.

Safety was drummed into the recruits constantly while on the firing line. Letting the rifle drift in any direction from the vertical position and pointing at the targets would result in kicks, blows, slaps, or being relegated to the Marksman Training Platoon (MTP) to be retrained in safety and other facets of marksmanship.

When we were on the 200 yard line we first fired 10 rounds to get our elevation and windage set for the day. These 10 rounds didn't count in the total of fifty rounds we would fire while on the course. After setting the "clicks" on our rifles we were ready for action.

Even though our daily scores were for practice only I was as nervous as a cat on a hot tin roof as I aimed for my first shot. I nervously got into the sitting position with one end of the rifle sling tightly fastened around my arm and sighted my M14 on the 6"bullseye of the Able target. I had twenty minutes to fire the first fifteen rounds on the Able target, 5 sitting, 5 kneeling, and 5 standing. Beside me was my "Data" book in which all my hits (or misses) will be charted by myself and my Marksmanship Instructor who was by my side as I fired the course. This would be the routine throughout this week while firing the course, including Friday, Qualification Day.

At the end of the first day I felt pretty good about my first day's score as I fired a 197 for the course. This score is a passing score and rates a marksman badge if fired on Qualification day. It felt good to achieve that score on the first day but I had three more days of practice ahead of me before Friday and Qualification day arrived.

The firing from the 500 yard line was as difficult as I had imagined. I had a problem with the vision in my right eye while aiming at the Baker target from the prone position from that distance. I was told that my vision was 20/20 during the eye exam but for some reason I had difficulty with a blurry vision at times from there. The fact that the targets were five football fields away might have been the reason. Despite the problem I still managed a couple of bull's-eyes from that distance. All in all I felt pretty good after my first day's work on the Rifle Range.

On Tuesday I had another good day in which I managed to fire a score of 196 which while not a great score was enough to achieve a qualifying score. The worst part of the day was that at noon chow we were once again fed bologna and salami sandwiches in the field. An apple and the accompanying pint of milk was my noon meal this day.

The heat and humidity continued to be a problem for everyone whether we were pulling butts or firing from the firing line. We were pleased to discover that the sand fleas weren't as bad on the Rifle Range as they were at our 3rd Battalion training areas but the mosquitoes more than made up for the absence of the sand fleas. Throughout the day we endured bite after bite from these insects and by evening we had ugly red welts all over our face, neck, and hands. Thankfully the long sleeves of our utilities protected our arms from being bitten.

Tuesday evening we double timed to chow as Private Christmas sang a beautiful marching tune that we kept in step with as we ran the two miles or so to chow. His singing was spontaneous and surprising to all the recruits, as we had no clue he could even sing. He was every bit as good as the Drill Instructors in leading our platoon's running cadence.

A possible score of 250 on the Rifle Range would be achieved if a recruit scored bull's-eyes with all 50 rounds that are fired on Qualification Day. Although no recruit has ever achieved a perfect score, several recruits from every platoon usually fire Expert on the Range. The table below shows the three Shooting Badges for different levels of qualification.

Marksman- 190-209

Sharpshooter-210-219

Expert- 220-250

Figure 1 Keeping score on the Range. USMC PHOTO

A shooting score below 190 on Qualification Day is an unqualified score and has many consequences for the unfortunate soul that fails to qualify. Promotions, job opportunities, and duty station assignments in the Marine Corps depend on a lot of different factors of which Rifle Range scores are of vital importance.

Wednesday started out as a routine day as my squad served their turn in the butts marking targets for the first relay of shooters. The day was cloudy and rain was threatening the whole time we were working in the pits that morning.

After chow my squad moved to the firing line at the 200 yard line. We had just finished firing our full 15 rounds there when the rain started falling heavily and the firing was postponed as everyone rushed for their field packs and grabbed their ponchos to put on.

As we were standing on the 200 yard line the Range Officer, a Lieutenant, shouted for the Drill Instructor to send a recruit to his two story range tower he was observing from. As I was standing near Cpl. Meade he shouted to me to report to the Lieutenant at the tower.

I ran to the tower and stood at attention as I saluted and shouted, "Sir! Private Nickels reporting as ordered, Sir!" The Lieutenant shouted to me as I stood at the base of the tower, "Private, go to the fire station and get my raincoat for me, and hurry back!"

I had no idea in the world where the fire station was but Privates at Parris Island obeyed orders and didn't ask questions when an order was given. I took off running (recruits ran wherever they went on Parris Island) along the road that ran alongside the range that was used to maneuver the mobile range tower between the 200,300, and 500 yard firing lines.

The rain was falling heavily as I ran to the main road that led to the Weapons Battalion as I knew I had observed numerous buildings in that area when on guard duty there and that was as good a place as any to start looking. Arriving at the Weapons Battalion area I found the fire station located just north of the Weapons Battalion Headquarters.

Reporting to a Staff Sgt. at a desk inside the fire station I explained that I was sent for the Range Officer's raincoat. Without speaking to the lowly Private (me) he retrieved the raincoat from a nearby office and I took off running again for the Rifle Range.

As I ran back up the road the rain had nearly stopped and I could hear the firing resume on the firing line. The range area was so large with its five separate firing lines that I didn't know exactly which range my platoon was firing on. I thought we were on the middle (Charley) Range but I wasn't sure so I went all the way along the main road to the first range and ran to the 200 yard line of that range. No firing was taking place on that range so I started running parallel to the 200 yard line to the next range. Again no recruits were where I had left them so I continued running.

As I ran I could hear firing from somewhere but I couldn't be sure where it was coming from. As I ran along the second (Dog) range 500 yard line I could hear the Zap, Zap, Zing of bullets striking the targets and I thought it strange that I could hear rounds striking the targets somewhere.

Then I heard a loudspeaker from somewhere barking, "Cease firing, cease firing on the firing line!" I looked to my left towards the 500 yard line of the Range and sure enough I could make out some recruits standing there. I finally came to my senses and realized I was running along an active firing line!

I quickly did an about face and streaked back the way I had come in and circled back to the main road. I then travelled down the range side road to the 500 yard line where I saw the guidon of my platoon flying. I reported to the tower with the raincoat I had been sent after and was questioned by the Lieutenant asking if I was the recruit that had just about been killed on the firing line. I denied everything and calmly played the innocent bystander routine. I don't think the Lieutenant believed me but since I had just come from the direction of the main road he gave me the benefit of the doubt.

I quickly retrieved my rifle I had left with the Drill Instructor and assumed the firing position on the line. My Marksmanship Instructor cautioned me to, "Be aware that some idiot was seen walking along the 200 yard line earlier and might try it again!" As I assumed the prone position and looked through the peep sight of my rifle I was sure that at least one recruit on the line wouldn't "try it again!"

As if my close call with the firing line wasn't enough trouble for one day, I had the unfortunate experience of not scoring high enough that day to qualify with a passing score. My final score was 186 which was four points below the minimum. The first two days on the line I had managed to qualify as a marksman but not this day. Although scores for the first four days of the week wouldn't be in our records, low scores could have undesirable consequences.

Beginning the first day of practice those who failed to qualify on the range had to report to the recreation room in our barracks at 2100 hours. That first night I watched as more than thirty recruits lined up at the door leading to the recreation room and awaited the appearance of the Drill Instructor from his "shack". No one knew for sure what would happen next but we all had a pretty good idea.

As the Drill Instructor made his appearance he stopped just outside the hatch and made a show of slowly pulling a pair of black leather gloves onto his hands. After pulling them on he made a fist with his right hand and slammed it into his left hand several times as we all watched quietly. He then opened the hatch into the recreation room and ordered the waiting unqualified recruits into the room. He then closed the hatch behind him.

When the hatch opened thirty minutes or so later the recruits came rushing out of the room, some bending over and clutching their midsections, some were holding their jaws, and others looking as if they had been roughly handled. We then knew that although our daily scores wouldn't count in the books, they would count for a lot in the eyes of the Drill Instructors.

The next night the unqualified recruits (unks) in the platoon for that day again lined up at the recreation hatch at 2100 hours. Again after a few minutes inside with the Drill Instructor they filed out looking a little disheveled and bewildered. Some of them were the same ones that had gone to the recreation room the night before.

I had seen the results of the previous two nights so I was under no illusions as to what I was in store for at 2100 hours this night. A few minutes before the magic hour the Drill Instructor made a show once more of pulling on his by now dreaded black leather gloves and heading for the recreation room. As he neared the hatch he looked over his shoulder and shouted, "Listen up! All my unqualified "Commandoes" form a line on the squad bay deck and march into the recreation room!"

I was nervous as I fell into line with the other unqualified recruits from the day's firing and wondered if my punishment would be to the face or to the stomach. I was a little worried because of the closeness of my stomach to my backbone but I was satisfied that whichever part of my body was destined to be pummeled this night I would take it like a man. After all I had endured everything that had been thrown my way the seven weeks I had been on Parris Island. I was now used to the physical and mental punishment that every recruit endured during boot camp and had weathered them all. This soon would pass, was my thought as we filed into the small recreation room with much apprehension.

As the last recruit came through the hatch, the Drill Instructor, who was waiting inside with one of his gloved fists pounding his other hand, shouted "Close the hatch and douse the light." When this was done he stepped to the far bulkhead and called the first recruit to him. Some light was filtering through the two windows in the recreation room and I could see the Drill Instructor swinging at the stomach of the recruit. One apparently powerful blow was delivered to the midsection, then he called for the next recruit to step up as the first recruit went to the back of the line and prepared for another round.

There were eight or nine recruits ahead of me but it soon came my turn. As I stepped towards the Drill Instructor I have to admit that my poor stomach was churning as I dreaded being punched in the very area that had given me the most trouble since I had been in boot camp. As I stepped in front of him at attention I tightened my stomach muscles to help me absorb the blow and breathed in a deep breath of air and held it to await the shock. In the dim light I saw his gloved fist swinging my way and I tightened my stomach even more to help soften the blow. As his glove touched me I involuntarily leaned forward slightly and grimaced as I waited for the pain to rack my body.

To my complete surprise I never had any pain! He had pulled his punch. Just before his gloved hand hit my stomach he had stopped the progress of his blow! I thought at first he had swung wide but then he called, "Next!" and I went to the back of the line. We all went through the line again and I thought sure I would get punched this time, but no, he again pulled his punch. What was going on here, I thought. Was this all for show or was he giving me a break. After all I had never given anyone any trouble during my training and had always obeyed orders. On the other hand, it was doubtful that he could tell who an individual recruit was because of the darkened room we were in.   Maybe the whole exercise was a sham just to scare everyone up a little. I never asked or found out what the other recruits experienced in the recreation room and I don't know to this day.

The next day was Thursday, Pre- Qualification day and by now we were used to the awkward positions we had to assume on the firing line of the Range. While today's score didn't mean anything as far as the record was concerned I was anxious to redeem myself because of yesterday's falter of not having a qualifying score. To not qualify today would cause a lot of worry on me when firing for the record.

Although the day was scorching hot and the mosquitoes were doing their thing all day long, I fired a score of 197 this day. I was pleased with my score and especially pleased that I wouldn't have to make another trip to the recreation room tonight. I didn't want to take a chance that the incident of the phantom punch might be real the second time around.

Friday May, 6, was qualification day and every recruit in our series of three platoons was plenty nervous and apprehensive about having a good score today. This score would go into our record books and would follow us throughout our time in the Marine Corps.

The Drill Instructors ordered each of us to attach one of our clothes pins with our name on it to the guidon. If we qualified today we were instructed to remove our clothes pins from the guidon. Any pins left on the guidon after firing on the range would clearly identify the failures to qualify. I advanced to the firing line hoping that I would be able to claim my clothes pin after firing the course.

After firing on the 200 and 300 yard lines I felt fairly good about my shots so far but I didn't know what my score looked like yet. We moved back to the 500 yard line and as the Marksmanship Instructor looked at my Data Book he remarked, "You need to score all fives from here to be sure you qualify today!" I didn't know if he was serious or if it was a motivating ploy but I was going to do my best either way as far as I was concerned.

I fired my ten allotted rounds and was worried because I only had three fives from this distance. The rest of the rounds hit in the fours area of the target. My stomach was in knots as the PMI added up the total score in my Data Book.

AS he finished adding my score he looked at me and with no emotion said, "Well Recruit, looks like you qualify as a Marksman in the United States Marine Corps. You scored a 194 today!" I couldn't help breaking into a big grin for the first time since I had arrived at Parris Island. Even the PMI had a slight grin on his face. Now I knew for sure that I had this Parris Island thing all but whipped and barring a major failure I would graduate from training in a couple of weeks!

After learning my score I proceeded to our platoon guidon and informed the Drill Instructor that I had come after my clothes pin. Several other recruits were also asking for their pins. Every time a pin was retrieved from our flag the Drill Instructor would shout, "Good job, Recruit!"

Qualification with the M14 rifle was critical to every Marine Recruit's career as I have stated elsewhere. It was likely the first thing a Commanding Officer looked at when going over a Marine's Service Record Book (SRB). Every Marine, with few exceptions, had to qualify with the rifle each year they were in the Corps. After all, the fact that we were rifleman first and any other MOS we might be assigned was secondary was drummed into us from the beginning of Boot Camp.

The past two weeks had been stressful and nerve racking for all the recruits on the Rifle range and we were very glad to see our time at the range coming to a close. The few unqualified recruits in our platoon had more work to do before graduation from the Island could be achieved. No more trips to the recreation room would be required after coming off the Rifle Range today. I don't think I, or any other of my fellow recruits was at all sorry to see those trips come to an end.

Qualifying on the Rifle Range was (and is) so vitally important to a Marine's career that the sometimes harsh methods used by Drill Instructors and Marksmanship Instructors to motivate the recruits can easily be excused by the results achieved. None of us felt any animosity towards any of the Instructors because of a few cuffs here and there. Most of us were just pleased to be able to pass this obstacle and go on to the next one.

Chapter Fifteen

FINAL PREPARATIONS

Figure 1 Company Commander's Inspection of Platoon 381- May 1966
(USMC PHOTO/Albert Love Enterprises)

Now that the pressure and stress of Rifle Range activities were behind us the whole platoon felt almost light-hearted as we tumbled out of our racks Saturday morning May 7, 1966.

When we sang the Marine Corps Hymn as we stood at attention that morning we made the squad bay reverberate with the sound of over one hundred voices shouting at the top of our lungs.

After our usual ten minute head call for each side of the squad bay we were ordered to unlock our rifles from the gun racks and prepare our field packs for light marching order. This meant some serious training for the day. The time spent at the Rifle Range had somewhat curtailed our regular training schedule so that we could concentrate fully on our shooting technique. We would today have a full day of drill and PT ahead of us.

We were in formation by 0500 and began a five mile conditioning run around the Rifle Range. I discovered my stamina had suffered a little the two weeks we were involved in Rifle Range activities. I was gasping for breath by the time we completed our run 45 or 50 minutes later. I wasn't by myself though. Every other recruit was breathing hard also. By the following weekend we were back to our old stamina levels.

We were now fully fledged Third Phase recruits and our platoon had the swagger to go with it. Our marching and manual of arms were now performed flawlessly by the platoon and our deeply suntanned faces, hands, and arms marked us as a veteran group to newer platoons.

When we reached a crossroad at the same time as a newer platoon we were given the right-of-way through the intersection first. We had done the same for older platoons when we first started training. As we marched by, the newer platoon's Drill Instructors would shout to their waiting recruits, "Take notice of how a platoon that knows what they are doing marches by! You miserable shitbirds will never be able to match them!" Those words made us very proud of what we had accomplished over the past weeks. Of course these newer recruits would feel the same as we did six or seven weeks from now after hundreds of hours on the drill field will have caused their drill evolutions to be flawless also.

After a full day of physical activities our platoon headed back to the Rifle Range mess hall and evening chow. After chow we continued to march and drill around the area for the next two hours. While we were marching near our barracks I was surprised to see a car slowly traveling up Wake Boulevard and looking closely at the several marching platoons. The occupants were obviously looking for a particular platoon, as they would slow down when they neared each platoon's guidon which had the platoon number emblazoned on it. When the car reached the head of our platoon I saw that it was my own immediate family in the car!

The windows were down and I clearly heard Mom say "That's Eddie's platoon! The flag had 381 on it!" My little sister Kathy shouted, "There's Eddie!" as she pointed towards my position in the third squad. I wasn't able to look at them or acknowledge them in any way as we were marching by.

They followed our platoon for an hour or so as we drilled and performed the manual of arms along the streets of the Rifle Range. I was a little embarrassed and hoped the Drill Instructor didn't notice the small Buick car following our every evolution of drill that we performed. When we stopped, the car stopped.

When we resumed our marching, the car followed. I can just hear my Mom saying, "Ed, keep up with them, don't let'em out of your sight. I can see Eddie there in the third row. He's the fourth one back!"

Wanda had written me a letter that they might try to visit me while we were on the Rifle Range but I never really expected that they would. I had written Wanda that I would only be in Weapons Battalion two weeks and I never knew where I might be on the Range. They had come anyway and I was glad they had of course. It had only been two or three weeks since the last visit but a visit by members of the family is always a great morale builder to a service member.

As we marched to our barracks and headed inside I saw the Buick parked nearby. The family watched as we ran up the ladder to our second floor squad bay. I was feeling good because I knew I would have visitors during visiting hours from 0800 to 1200 hours tomorrow. I could hardly wait to see Wanda and the rest of the family.

As we assumed the position of attention in front of our racks the Drill Instructor, Sgt. Shue shouted, "Do any of you knot heads know anything about the red car that's been following us all evening?" I swallowed hard as I took one step towards the middle of the squad bay, clicked the heels of my boots together, and shouted loudly, "Sir! This recruit does sir!" He then turned on his heel and shouted over his shoulder, "Get the hell in my office Recruit!"

Uh oh, I thought. I hadn't counted on the wrath of the Drill Instructor coming down on me. I screamed loudly, "Aye Aye Sir", as I quickly ran to his office hatch and assumed the position of attention beside the hatch. I then stood beside the outside frame of the hatch and slapped the frame three times with the flat of my hand. He shouted, "Enter Recruit!" As I stepped forward one step I did a left face and entered his office. "Shut the hatch Recruit!" He shouted. I quickly shut the hatch behind me and stood in front of his desk as I shouted, Sir! Private Nickels reporting as ordered, Sir!"

Looking at me with a scowl on his face he said in a soft tone, "Private is your family in that car?" Staring straight ahead I shouted, "SIR, yes Sir!" Then he said, "Is your wife in that car also?" Again I replied, Sir, yes Sir!" Still talking in a low tone of voice he said, "Recruit, I'm ordering you to go to your rack, get your cover, and hurry outside to see if you can catch your family before they leave. It's now 2010 hours. Be back here at 2145. Do you understand me?" I was almost too dumbfounded to reply to him. I had never seen a recruit catch a break like this. It must be a test or a trap, I thought. I hesitated a few seconds as I tried to figure out if he was serious or not. He looked at me and said , "Private, I gave you a direct order to visit with your family, now get out of here!" "Aye Aye Sir, I shouted. I took one step backward, did an about face, then took off running for my rack to get my cover, and went out the back exit and down the ladder.

When I reached the bottom deck I saw the Buick still parked where I had last seen it and took off running towards the car before they could pull out. I went to the back door where I could see Wanda looking out the window of the car. She quickly opened the door and got out of the car. We stood there hugging without speaking for several minutes as Mom was softly crying and saying," You're so skinny and tanned. Are you eating enough?" I assured her I was. They were all as surprised as I was to be able to see them at all this evening. Mom said,

"How did you get away to see us?" I answered, "I have no idea. I guess my Drill Instructor has a heart after all."

We sat In the parklng lot and talked, then drove around for awhile before I headed back to the barracks. I enjoyed seeing Dad, Mom, Wanda, Kathy, Rita, Jimmy, and Phil that Saturday evening in May.

The next morning I again got to visit with them for four hours before they headed back home. Graduation was just two weeks away and they planned to try to attend that ceremony if possible. I was in very good spirits as they headed back home Sunday evening. I was looking forward to their next visit in two weeks.

The generosity shown to me that evening by Sgt. Shue was much appreciated and I always wanted to tell him just how much I appreciated it but I was never able to do so. By allowing me an extra few hours with my family he was going against visiting regulations and could have easily got into trouble himself.   Again, I thank you, Sgt. J.C. Shue, USMC. Your action renewed my faith in humanity that evening in May, 1966, just when I had begun to wonder if there was any humanity left for lowly Marine Recruits at Parris Island.

Sunday afternoon at 1300 we saddled up with our fully loaded field packs, haversack, rolled up poncho, rifle, bayonet, two 7.62mm magazines, first aid kit, full canteen, helmet, helmet liner, and strapped our fully overloaded seabags on our shoulders as we started our conditioning march back to our 3rd

Battalion barracks. I had grown to like living at Weapons Battalion and the bad memories of the two weeks of indoctrination and "attitude adjustment" our platoon experienced in our 3rd Battalion squad bay had left a bad taste in my mouth concerning that place.

The only consolation to moving back was the knowledge that we only had two more weeks of training left on Parris Island before we were scheduled to transfer to Camp Geiger, North Carolina where we would undergo further Infantry training at Infantry Training Regiment, (ITR). We were told that the training at ITR would be even more intensive and demanding than what we were experiencing at Parris Island. After the experiences we had undergone on Parris Island, I shuttered to think about how difficult ITR must be.

We were loaded so heavily as we marched through the fields and under the tall pines that it was almost impossible to double time on our hike back to 3rd battalion. We marched so rapidly in the heat that every man was struggling and gasping for breath when we arrived at our 3rd Battalion barracks area.

All four of our Drill Instructors made the conditioning hike with us and they were unimpressed with our effort and unsympathetic with the heat difficulties we experienced during the hike. They said we were "worse than a bunch of old women" when it came to being able to withstand the heat while hiking with a heavy load. They assured us they would "run us back into shape over the next two weeks." (And so they did.)

On Monday, May 9, we had reveille at 0330 and ran the Obstacle Course before we went to the mess hall for chow at 0600. After going through the course once we turned around and ran it again. Having had an absence of two weeks since we had tackled the course I had nearly forgotten how tough the course could be. It was going to take a little while to get back into the groove of things, it seemed.

After morning chow we were marched to the base barber shop for a haircut. This would be our third trip to the barber shop in six weeks. We were surprised to learn that this haircut would be a "high and tight" cut instead of being sheared completely bald. We were all left with a strip of hair running down the middle of our heads, resembling a Mohawk cut. The high and tight haircut was another symbol of our Senior Recruit status. The sides and backs of our heads were once again shorn to the skin. This would be our last haircut before graduation would occur.

That same afternoon we marched to a large building and received our final clothing issue. We were issued the winter and summer uniforms we would wear as Marines. We were issued a set of greens, browns, and khaki shirts and trousers . The greens were for winter wear and the tropical and khaki uniforms were for summer. The tropical uniform was the more formal uniform for summer use. No Dress Blues were issued as this uniform had to be purchased at our convenience and at our own expense after graduation. We were not yet considered Marines and would not be eligible to wear Dress Blues until graduation from boot camp. I couldn't afford them at that time anyway.

The process of issuing, fitting and performing alterations on each recruit's uniforms took the whole afternoon to complete.

As each item was issued we folded them and placed them into our seabags which we had brought with us. We would later in the week send our uniforms to the base laundry for cleaning and pressing to prepare for our wearing them on graduation day.

We returned to the 3rd Battalion barracks at 1700 and ran the Confidence Course before we headed to chow. The Confidence Course requires a recruit's maximum effort both physically and mentally to complete.

The eleven obstacles of the course demands speed, agility, upper body strength, endurance, and disregard of heights. "[1]

We would have to run the Obstacle Course and Confidence Course nearly every day for the next two weeks. The Confidence Course lived up to its name by increasing our confidence of being able to overcome nearly any obstacle we were confronted with during boot camp. Obstacles that we had considered impossible to overcome became commonplace to conquer after doing them over and over throughout our training.

Monday evening at 2000 we were engaged in our normal routine of standing in front of our racks at the position of attention when the Senior Drill Instructor, S/Sgt. Lawson called the names of seven recruits and ordered them into his office. One of the recruits he called was Private Moore, one of the friends I had made several weeks ago while we were on Maintenance and Mess duty.

The recruits returned to the squad bay after an hour or so behind the closed hatch with S/Sgt. Lawson. As our "free time"

---

[1] Da Cruz, Daniel, *Boot*

had started at 2100 I was able to ask Private Moore if he wanted to discuss what went on inside the Drill Instructors' shack after he reported there. He grinned and said, "They wanted me to apply for Officers Candidate School," (OCS). I then asked, "What did you tell them?" " I told them thanks, but no thanks," he said in a lowered voice. He then told me that he was satisfied with his career as a school teacher and he hoped to return to his job after his service obligation was over. He continued, "With my luck I'll probably have a 0300 MOS, (infantry), and go straight to Vietnam anyway. The government loves to make college boys Infantrymen. I'd rather serve as an enlisted man if I'm assigned to the Infantry, so I have no desire to be an officer. A couple of the college guys that were interviewed with me said they wanted to apply for an OCS slot. It's not for me though."

I respected his decision but I told him that I probably would have jumped at the chance if I had been approached. The Marine Corps wasn't in the habit of asking high school dropouts to be officers though. I had an adequate General Classification Score (GCT) to be an officer but I lacked the critical qualification of a minimum of two years of college to be considered. It was another rude awakening for me to realize that my lack of education would probably always hinder my chances of advancement in the real world.

That same evening Private Wilson, the other friend I had made while on Maintenance and Mess duty, was called to the Company Office to take an incoming emergency phone call. When he returned he came to Pvt. Moore and me and told us that the phone call was from his wife, informing him that she had given birth to their baby, a boy, that evening. He was ecstatic and informed Moore and me that he would buy us a cigar, "If I get out of Parris Island alive," as he put it.

Another unfortunate incident occurred during this week that started innocently enough but could easily have ended with dire consequences for those involved. We were standing at attention in platoon formation while two of our Drill Instructors were conferring with one another a few yards away. Two recruits were carrying on a whispering conversation between themselves, thinking that they couldn't be overheard by the Drill Instructors. They were wrong. One of the Drill Instructors called them down and threatened the two recruits with PT if their whispering continued.

The two recruits stopped their conversation after the chastisement by the Drill Instructor but after a minute or two they resumed their whispering. The Drill Instructor that had threatened them with PT again overheard them and called out one of the recruits by name. With seething anger the Drill Instructor shouted, "Private T...., report to my office when we get back into the barracks!"

As we filed into the barracks a little while later the still angry Drill Instructor headed straight to his office. As we stood at the usual position of attention the Drill Instructor stuck his head out the hatch of his office and called out for the guilty recruit to "report to my office on the double!" The recruit ran into the office and the hatch slammed behind him as he entered.

The Senior Drill Instructor, S/Sgt. Lawson was in the squad bay with the rest of the platoon and was talking about the upcoming inspections the platoon would be standing prior to graduation the following week. Suddenly we heard a series of loud shouts, crashes, and sounds of items falling to the deck coming from inside the office where the Drill Instructor was confronting the recruit. S/Sgt. Lawson stopped his instruction as he listened to the loud slams coming from his office. We recruits stood in the squad bay in front of our racks with our own thoughts and ideas about what was happening inside that closed hatch of the Drill Instructor's shack.

The whole squad bay was so quiet that you could have literally heard a pin drop as we assumed an intense battle was happening between the Recruit and the Drill Instructor. After about five minutes had elapsed with no abatement of the racket the Senior Drill Instructor, S/Sgt. Lawson, ran to the office hatch, opened it, and went inside and closed the hatch behind him.

A few moments later the hatch opened and Private H. ran back into the squad bay. As he passed in front of me he looked as if he had been through a rough time. The impression I got as I watched him pass by me was that he had been in an altercation of some sort but I saw no visible signs of combat.

In a few minutes S/Sgt. Lawson came out to the squad bay and made a speech about how it was regretful that we recruits still hadn't learned to obey orders even at this late stage of our training and that he was disappointed that harsh measures were still required to enforce good discipline and order in the platoon. He said something to the effect that he hoped another incident like this one never occurred again.

I have often reflected back on this incident over the years and I have since come to the conclusion that this whole incident was a sham. The recruit involved in the incident was not known to be a troublemaker and had never, to my knowledge, been in any kind of trouble before during boot camp. As a matter of fact that particular Recruit had been the high scorer for our platoon at the rifle range and enjoyed the respect of the Drill Instructors because of that accomplishment. It was not likely that he would have been picked out for an especially harsh punishment.

The Drill Instructors could not let an incident such as talking in ranks go unpunished because of the danger of other recruits taking advantage of it and leading to poor discipline in the platoon. Our training had made us into recruits that obeyed orders instantly no matter the consequences of life or limb. This was what was required of all Marines in the Fleet. If you were ordered to "attack that hill," you were expected to do just that and to do it instantly. One incident of poor discipline could spoil the whole platoon in boot camp where corporal punishment was swift and sure for disobeying orders.

That's why I think that the whole fight in the Drill Instructor's shack might have been staged for the good of the whole platoon. We all *thought* there was a fight going on in the office and whether it did happen or not it caused our platoon to be on our best behavior during the last two weeks of boot camp. Nobody wanted to lose their teeth or suffer a broken jaw in the final stages of our ordeal of conquering boot camp.

Tuesday morning we were on the drill field and going through the manual of arms with the confidence and knowledge that we could compete with any platoon or any Ceremonial Drill Team when it came to the evolutions of drill. When the Drill Instructor at one point shouted, "Left shoulder, Arms," we shifted our M14's from our right shoulder towards our left and we heard a loud "crack" as a rifle stock shattered into pieces! The nightly slapping of our palms against the concrete deck of our barracks had finally resulted in a broken rifle stock when going from right shoulder to left shoulder arms!

The Drill Instructor shouted a loud," Platoon Halt!"He called the recruit that achieved the distinction of a broken rifle stock to the front of the platoon and ordered him to go find a seat in the shade somewhere and rest while the rest of the platoon continued drilling! We were flabbergasted. Whoever heard of a Marine Recruit getting such a break while at Parris Island? It was obvious that there were benefits for achieving this unusual goal. As we resumed our drilling the slaps of our hands hitting the stocks were louder than I had ever heard on the drill field. Every one of us recruits, (including me of course,) was almost breaking the bones of our hands trying to shatter our rifle stocks. Even several officers passing by the 3rd Battalion drill field stopped to watch and listen as we made the wood of those rifle stocks ring loudly that Tuesday.

In just a few minutes we heard another "craaccckkk" as another recruit achieved the almost impossible goal of shattering his rifle stock. Bystanders were applauding as the proud Drill Instructor once more called for the recruit with the broken stock hanging at the end of his rifle sling to come forward to face the platoon. The Drill Instructor excused him to "find a place to relax out of the sun somewhere!"

Try as I might to break my own rifle stock, all I succeeded in doing was to make the palm of my right hand look like a piece of hamburger meat by the end of the day. By the end of the week five or six other recruits of our platoon succeeded in breaking their stocks and were rewarded with time off training for a short period. The Drill Instructors were so thrilled with our marching and drilling that we felt a noticeable reduction of stress and mental pressure that week.

Later in the week we had a conditioning hike to Elliott's Beach which was located beyond the Weapons Battalion area near Page Field, the then (and now,) unused Marine airfield located on Parris Island. We once again loaded our field packs with our mess gear and other items needed for our bivouac at Elliott's Beach and rolled our haversacks and ponchos up for carrying to the field. The rolled ponchos were attached to the back of our webbed cartridge belts and the haversacks went atop our field packs. The heavy helmets and liner would be worn during our march and would make our misery complete that hot day in May.

Arriving at the Elliott's Beach Training area we paired off and set up our pup tents by having two recruits getting together to fasten their shelter halves together to make a two man tent. After we got our campsite set up we were directed to where testing sites were set up and manned by the Drill Instructors of our 380,381, and 382 recruit series of platoons. We were then tested orally and by practical application on various procedures and subjects.

The knowledge of the characteristics and capabilities of the M14 rifle was one of the tests. Other tests involved the Code of Conduct, First Aid, the eleven General Orders, Customs and Courtesies, the Uniform Code of Military Justice, (USMJ) Marine Corps History, and many other subjects we had studied over the past eight or nine weeks. We worked well into the night and didn't hit the racks, (pup tents) until 2300 after a long and tiring day.

Reveille was sounded at 0300 and we had a quick breakfast of c-rations and canteen water. Spaghetti and meatballs for breakfast wouldn't have been my first choice. Neither were the lima beans my tent mate tried to trade for my spaghetti and meatballs. I declined his offer, much to his chagrin.

We again resumed our testing of our Marine Corps knowledge with a couple hours respite after noon chow to PT and to run a three mile conditioning run with packs, rifles, and helmets. Several recruits were nearly in the stage of heat exhaustion after the run, due to the humidity and heat.

The majority of the recruits had become serial cursers and black guarding regulars by now and I'm sorry to say that I had somewhat of a habit myself of uttering nasty words after being around so many practitioners of those dirty words. I declined and abhorred the use of swear words though. The use of swear words intermixed with the F word was so commonplace and so rampant that I feared that while eating a meal with the family gathered around the table at home I might blurt out a dirty word or two. Shouting out "Pass the f...ing peas! "most likely wouldn't have passed muster in polite company.

One evening at 2130 we were sitting on our buckets and cleaning our rifles when a recruit screamed out, "Spy on deck!" This was the signal that a stranger other than an officer was inside our squad bay. Glancing up I saw the tallest and largest recruit I had seen on the Island standing at attention in front of our Drill Instructor, S/Sgt. Lawson.

Before he could state the reason for his presence S/Sgt. Lawson ordered him to drop to the deck and give him 25 pushups. Then he had him to crawl along in front of us recruits all around the squad bay and "bark like the Dog that he was," as the Drill Instructor stated it. He crawled around on all-fours barking and scratching the deck with his paws, (fingers), and howling to the top of his lungs. After about five minutes he was ordered to his feet. He stood at attention, towering over the diminutive S/Sgt. Lawson as he shouted that he was the company runner and had come with a message from the Company Commander.

S/Sgt. Lawson read the message the large recruit handed him then said, "Recruit, what is your name?" The stranger shouted out his name which elicited a low gasp from the other recruits of my platoon that were within hearing distance of his answer. It was the name of a San Diego Chargers professional football player, according to those recruits, as they later related to me and others of the platoon.

Whoever he was, S/Sgt. Lawson wasn't overly impressed and ordered him to assume the position of a barking dog and to keep it up as he went out the hatch. I watched as this large man dropped to his all-fours and scrambled through the back exit hatch, barking and howling for all he was worth. A position of fame, status, or national popularity has no place for mercy when it comes to the relationship between Drill Instructors and lowly U.S. Marine Corps Recruits.

Saturday, May 14 was a very important in the minds of every recruit in my platoon. This was the day our series of platoons would take our last Physical Readiness Test. (The dreaded PRT.)

We had been practicing running the PRT for weeks at every available opportunity, even while at the Rifle Range. No matter how many times we ran it with the fully loaded field packs, the rifle, bayonet, the helmet, helmet liner, full canteen, poncho, haversack, web belt, first aid kit, and other accouterments, it never got any easier. Adding to our misery was the heavy combat boots and the unrelenting heat of the day.

It seems that the only time it rains and cools the atmosphere on Parris Island is when your own platoon is on the Rifle Range. The weather seems to be officially regulated so that the temperature is at its highest and the sun is at its zenith when the PRT is run! At least I found it to be so.

We were told over and over during training that Marines never, ever leave a fellow Marine behind for any reason, whether in combat or in training. If a recruit should fall out of a run or any other training exercise, the closest men picked up the stricken Marine and his gear and carried them over the finish line. As a platoon, we had already had to drag or carry a couple of recruits over the course when the heat of the day proved to be too much for them. Nobody was allowed to not finish the Obstacle Course, the Confidence Course, or the PRT. Death or serious injury was the only excuse for not finishing.

This may seem harsh but in fact, it was somewhat comforting to know that one could depend on their fellow recruits to see that they passed the tests no matter the difficulties involved. No recruit wanted to fail and be set back in training, especially at this stage of their torturous training.

We stood in platoon formation at 1300 and stood inspection by our full complement of four Drill Instructors to assure them that we had our full amount of impedimenta required for the

PRT. Already the heat and humidity was almost stifling. A more undesirable day for running couldn't have been wished for by the most sadistic of Parris Island Drill Instructors.

After the inspection revealed no major improprieties of battle dress we proceeded to march to the area of the test which would be run on a large circular dirt training area. The distance around the oval track was one half mile. It would take six trips to complete the required three miles. The three miles were to be run at full speed, with full combat equipment, and a time limit of 24 minutes.

We had run the course many times throughout boot camp and as stated before there had almost always been several men that either had heat exhaustion or been close to it during or after the run. Occasionally one or two men would collapse during the run and would have to be carried or helped over the finish line.

The run was only part of the total PRT requirements. The other events included a fireman's carry of another recruit for a distance of 150 yards, 62 step-ups onto and back off a 24 inch high concrete platform, a vertical rope climb of 20 feet to the top of a head log, and of course the three mile run in 24 minutes. All events were tackled while carrying a full combat load on our persons. Passing these requirements every three months was essential to being ready to deploy anywhere in the world at a moment's notice.

The first mile of the exercise was uneventful but shortly thereafter a recruit in the fourth squad, just to my starboard (right) side, stumbled and fell in a cloud of dust. I quickly ran to his side and picked up his rifle as several recruits picked the fallen recruit up and resumed their run, now at the rear of the platoon with the stricken recruit.

As I picked the rifle up I shouldered it on my left shoulder by the rifle sling and continued carrying my own rifle at "port arms." I had to increase my speed considerably to catch up with my squad and resume my place in the squad. Having already being nauseated by running in the heat of the day, I was now in complete misery as we ran in step around the field. Waves of nausea were washing over me and I was "dry" retching and feeling faint as we reached the two mile mark. I was always proud of having never fallen out of a run or any other physical exercise while at Parris Island but my clean record was in dire danger that day as we circled that dusty, unbearably hot, and seemingly endless dirt track that Saturday afternoon.

I did a whole lot of praying to be able to finish that run and I doubt I was the only one asking for Devine help to be able to finish successfully. When we at last were running the sixth circuit around the track I knew that I was going to make it without a PRT failure on my record. Thank you Lord!

When the run ended the whole platoon was bending over at the waist and coughing, throwing up, and generally near the point of heat exhaustion or even heat stroke. The recruit that had to be carried over the finish line had regained conscientiousness after he was carried over the line. He had suffered heat exhaustion as a result of the heat and humidity combined with the stress and physical exertion of the run.

We had run three miles and even five miles before while at Parris Island but the heat had never been as bad as on that particular day. No strenuous physical training was conducted during "Black Flag" days when the heat was over 95 degrees F.

On those days an actual black pennant was on prominent display at the training areas to advise the platoons to keep training exercises to a minimum. I think they missed the boat in not displaying the flag during our PRT run.

At the end of the day we had completed the other requirements of the PRT without any major problems. We were now ready to make final preparations for our final week of boot camp that would be completed the following week.

We were fairly confident now that the remaining recruits of our platoon would be able to hang on for our last week of training, although one could never be sure until that day actually arrived. The coming week would be rather hectic as we were scheduled to stand several inspections, prepare our equipment for turn in, receive our orders, and prepare for graduation exercises.

Sunday afternoon platoons 380,381, and 382 marched to the 3rd Battalion Parade deck and stood at "Order Arms" as the Company Commander, Captain Haley, prepared to inspect the three platoons of our series.

We were lined up in platoon formation of four squads of nearly 25 men in each squad. Our rifles had been cleaned for hours the night before, using cotton swabs to reach the deepest recesses of our M14's. After a thorough cleaning with soap and water,(I couldn't believe it either,) our rifles were wiped down with a light coat of oil. Before our PRT Saturday morning we cleaned them a couple more hours and after the PRT we cleaned them again. We were as ready as we would ever be for inspection.

As the Captain went from recruit to recruit, he was accompanied by several NCOs who carried notebooks to make notations that the Company Commander might dictate to them. The Captain would step in front of each recruit, make a right face as the recruit snapped to attention, brought his rifle to port arms, and released his grip on his rifle just as the captain grasped the rifle at the hand guard and began inspecting it for rust, carbon, cleanliness, and any other slight imperfection of the rifle. As he closely looked each rifle over he would ask each recruit a series of questions. As he stepped in front of me he asked, " Where are you from Private?" The next question was, " What's your second General Order Private?" I answered them to his satisfaction and he smartly returned my M14 to me after his quick inspection.

He then measured the distance between my "Overseas Cover" (also called Piss Cutters) and the top of the nose which regulations required to be precisely the width of two fingers. He also looked each Recruit's "Tropical" brown summer uniform over for pennants (hanging strings) as he completed each recruit's inspection of rifle and person.

Each recruit was rated as average, above average, or outstanding in their record books by the Company Commander. Overall our platoon and our series of three platoons had an "Outstanding" rating.

It felt good to get over the hurdle of this important inspection. We still had one more important inspection to stand, the Battalion Commander's Inspection, which was scheduled for Monday morning.

Chapter Sixteen

GRADUATION

After church services Sunday morning Drill Instructors S/Sgt. Lawson and Sgt. Shue called the platoon together in the middle of the squad bay in a School Circle. This was a simple formation where the recruits gathered around the Drill Instructor in a circle as we sat cross legged and listened to an important instructional class or message. Although the Circle was an informal affair, no talking, whispering, or interjection of opinions was allowed from the recruits unless solicited by a Drill Instructor.

The gist of the message that day was centered on what would be expected of us as we entered our final week of training and prepared for graduation from boot camp. We were reminded that boot camp was just the first phase of training. We still had several months of training to undergo after Parris Island before we would join the Fleet in our designated MOS fields. We were informed that we would learn what each recruit's MOS would be sometime during the coming week.

Sunday night I wrote my last letters home as a Parris Island Recruit. We would be too busy preparing for graduation to even write a letter at night, the Drill Instructor informed us. I wrote my Grandmother, Mom and Dad, and Wanda, and included a note to my little seven year old sister Kathy.

Hello Kathy, "How's my little sister? Did you enjoy your trip to see me? Be good until you go to school and help Mommy around the house. Kiss Wanda for me and tell Jimmy, Phil, and Rita hello.

Love, Eddie

P.S. I'm glad you know your bus number now!

Monday morning May 16, we marched to the 3rd Battalion concrete Parade Deck at 0645 to stand our final inspection, the Battalion Commander's Inspection. I was exceptionally nervous and filled with apprehension as we marched to the Parade Deck for the inspection. After all, the highest ranking officer in the battalion would be inspecting us and he had the final say in whether each recruit was ready for graduation or not.

Lt. Colonel Lillich arrived at the Parade Deck at exactly 0700 and started his inspection with platoon 380 which was aligned in platoon formation to our platoon's left. We were at "Parade Rest" with our left hand and arm resting on our lower backs as we held our grounded rifles by the muzzle with our right hands and awaited our turn to be inspected.

The inspection of Platoon 380 took twenty minutes or so, during which time I kept going over in my head the eleven General Orders and the nomenclature of the M14 rifle. We had been told that the Colonel generally asked one question concerning one of those two subjects and that we better know the answers when asked.

When the Colonel began inspecting our platoon he started with the fourth squad which was in the front row of recruits. After inspecting the fourth squad he came back down the line of recruits to our third squad leader and started inspecting him.

I was the fourth man to be inspected in our squad so my wait was only a couple of minutes before he stepped in front of me and executed a "right face" before me. I came to attention as I brought my rifle to "Inspection Arms," pushed the receiver lever of my rifle back with the heel of my right hand, glanced quickly down at the open receiver, then locked the receiver open by setting the locking lug with my right thumb.

As the Colonel's right hand touched the hand guard I let the rifle go and stood at attention as he inspected the rifle for rust, carbon deposits, or dirt. As he looked the rifle over he asked me, "What's the muzzle velocity of the M14 rifle, Private?" Without hesitation I shouted out, "SIR! The muzzle velocity of an M14 rifle is two thousand eight hundred feet per second, Sir!" "Very good, Private, keep up the good work!" With those words he held my rifle in front of me to grasp and I then proceeded to assume the position of order arms with a rifle salute as he did a left face and stepped in front of the next recruit.

A sense of euphoria came over me as the Colonel moved along the ranks of my platoon and finished his inspection. Knowing that I had successfully passed the final hurdle in our training schedule created a great feeling of relief from the constant fear that any failed training event could result in an extended training period or being set back.

That same afternoon our platoon marched to the base armory and turned in our M14 rifles to be used by another incoming new platoon of recruits. They had served us well during our time on the Island. They were always close by us during training and some recruits had even slept with their rifles during training. Sleeping with your rifle was reserved for those who accidently dropped their rifles to the deck. If a rifle hit the deck the recruit that dropped it automatically carried it to his rack just before lights out and laid it on the rack beside him. The duty fire watch recruit would check the offender's rack every two hours to make sure the rifle was still in the rack beside him.

After turning our rifles in we marched to a supply warehouse near the 2nd Battalion area where we turned in all our "782" gear, including our familiar galvanized buckets. The old pails had served us well as seats, scrub buckets, equipment carriers, and even as water containers, and were in every way an essential part of our equipment. We spent many hours of boot camp sitting on them as we cleaned our rifles, shined our shoes and boots, polished our brass belt buckles, sewed our torn uniforms with our "housewives," (sewing kits), and wrote our nightly letters. I would never look at galvanized buckets the same way again after boot camp. Let's just say that I left Parris Island indelibly impressed with the ring on the bottom of the galvanized bucket after using it as a seat for so many weeks.

Tuesday began with a five mile conditioning run at 0600. Running without our rifles and other accouterments felt awkward after so many weeks of running with them on our persons. The absence of all that extra weight was a relief that we were thankful for. We all enjoyed the run very much, especially since the heat and humidity was mostly tolerable this early in the morning.

The afternoon was taken up with running the Obstacle and Confidence Courses for the last time. We were able by now to run either or both of them without missing a beat and with much less pain and agony than when we first attempted them.

Those recruits that had entered boot camp overweight had by this time lost several pounds of weight. Some had lost as much as twenty pounds during training. Most of those that had arrived on the Island underweight had put on several pounds. I wrote a letter home in May saying that I had gained three pounds since arriving at Parris Island but I think I was trying to make the family think I was enjoying boot camp and even gaining weight. Of course I was fibbing. I had no way of determining if I was either gaining or losing weight. The Marine Corps had, for some reason or other, neglected to install weight scales in the barrack heads. (A small attempt at humor here.)We were not weighed at the end of boot camp but I appeared to have lost weight instead of gaining.

The last week of training I could do more than 80 sit-ups in two minutes or less. I had increased the number of pull-ups from the three I managed during the Initial Strength Test to the ten I could do during the last week. Our run times for the three mile PRT and the five mile Conditioning Runs had increased dramatically also. I was proud that I had never fallen out of a run of any distance even when I thought I couldn't take another step at times. I could now do 25 pushups with relative ease but still struggled while doing 50 at a time. I could easily do 500 side-straddle-hops in one session without tiring. I had proved this several times during boot camp after I had stirred the ire of a Drill Instructor along the way and was ordered to the quarterdeck for "Incentive Physical Training." (IPT)

Visits to the quarterdeck for IPT not only served as a punishment but a side benefit was that it built up the muscles while one worked off an infraction. I never minded the bends and thrusts, sit ups, and the side-straddle-hops, but I dreaded the pushups, especially when a count of fifty or more was ordered.

Tuesday evening at 2000 hours the Senior Drill Instructor, S/Sgt. Lawson, strode into the barracks squad bay with a clipboard and stood at the desk located in the near the quarterdeck area.

He informed us that he had a list that identified the MOS of each recruit and wanted us to listen up as our names were called. This was what most of us were waiting to find out ever since we received our draft notices from Uncle Sam. We all assumed we would be classified as Basic Riflemen or "Grunts" as Marines that carried the 0300 MOS were called. All of us had listed three choices of MOS when entering boot camp but we all assumed that all draftees would be riflemen and that we would be sent straight to Vietnam after training.

The Drill Instructors had pounded that fact into our heads over and over all through boot camp. "All you slime balls are going to Vietnam as Grunts and you better be ready to kill "Charlie" before he kills you, was a common refrain as we practiced bayonet tactics or fired on the rifle Range.

As S/Sgt. Lawson began calling the assigned MOS of each recruit by alphabetical order I was surprised that at least half of the assigned MOS classifications were those other than the 0300 Basic Rifleman MOS. Some received the 3300 MOS which was a Basic Food Service Man (cook), others received a 3500 MOS, or Basic Motor Transport Man (truck driver or mechanic).

Some other jobs assigned to the recruits whose names were called ahead of mine were classifications in artillery, tanks, communications, aviation, and combat engineers. I knew that an assignment in a Combat Arms MOS would almost certainly result in going straight to Vietnam after all our training was completed. If I received an assignment to one of these jobs I would not be able to keep my promise to my wife that we would eventually be able to rent a place near my duty station. I was on pins and needles as I waited for my name to be called.

My friend Private Moore's name was called before mine and he received a 0300 Basic Rifleman MOS. Since he had a Bachelors Degree and was a school teacher I was surprised at his assignment. If he was being made a Grunt I figured I would surely be a Basic Rifleman also.

S/Sgt. Lawson soon called my name and glanced at me as he called out, " Private Nickels, Eddie G., 0100 Basic Administrative Man!" I was almost flabbergasted to learn that I was tapped as an office clerk of some sort. The job descriptions of this MOS included Personnel Clerk, Unit Diary Clerk, Administrative Clerk, and Postal Clerk. Who would have thought that I would end up with the job that I had listed as my first choice during our testing in Receiving? My good friend, Private Moore, a college graduate, was assigned as a rifleman and here I was, a high school dropout, receiving an office job. I could hardly believe my luck. The ways of the military hierarchy were hard to understand at times.

If the averages held true for our platoon at least twenty five percent of the members of the platoon received the Basic Rifleman MOS. Both of my best friends, Privates Moore and Wilson, were classified as Basic Riflemen. When we were later able to discuss our assignments among ourselves Private Moore expressed his opinion to us as to why so many of the college graduates of our platoon were classified as Riflemen. He said that, in his opinion, the Marine Corps probably assumed that the college men would serve their military obligation and would then leave the service, while non-college men would choose to stay in the service as a vocation. At the time his opinion sounded reasonable to me but over the years I have discounted his theory. I now think that the Marine Corps assigned us where we were needed but granted each recruits choice of job as far as possible. Regardless of the MOS assigned to us, the Marine Corps still considered us Riflemen first. Even most Marine fighter pilots spend some time with Marine Infantry organizations, either as forward observers or as leaders at the Platoon, Company, Battalion, Regiment, or Division level.

When enlisted Marines are in the area where battles are being fought they are expected to be able to pick up a rifle or other weapon and join with the Infantry Marines to repel attacks or to charge the enemy. They have had to join the Grunts in the foxholes, in the trenches, and join in the attack many times in all the wars fought by the Unites States.

The Vietnam War had no front lines and today's War on Terror is being fought much the same way as the Vietnam War. There are many recorded incidents where cooks and truck drivers in these wars have picked up rifles and joined a desperate battle for survival. This is why all Marines are required to undergo several weeks of Infantry Training no matter what their regular job might be.

Our platoon marched to the Main Parade Deck on Wednesday and Thursday afternoon to go through graduation rehearsals. It's very important to get the cadence and the different commands just right because of all the visitors and military brass that attend these events on Parris Island. Besides, none of us wanted to make a mistake in front of our families while they sat watching us from the reviewing stand.

Wanda and some of my immediate family were planning to be there Friday for the ceremony but I would not be able to see them before the graduation ceremony was completed. I was hoping they would be able to make it but wouldn't know for sure until Friday. No messages or phone calls were supposed to be allowed to be made or be received by recruits while on Parris Island, although individual Drill Instructors would sometimes bend this particular rule in special circumstances.

A total of ninety six recruits would be graduating from my platoon #381 on Friday. One hundred and ten tired, scared, and confused recruits had stood in front of the racks inside the 3rd Battalion Barracks on the day our Drill Instructors ran us all the way from Receiving. We had endured at least a week of "Indoctrination" in which we had been broken down completely, both physically and psychologically. During that period each recruit had been made to feel that a group of psychos had control over us and that we might never get out of this place alive. After we had been sufficiently broken we had been slowly built back to the point where we could hold our heads up and be proud of what we had accomplished.

Some in my platoon swore that they would "get the word out" as to our treatment during boot camp. All that was forgotten as our graduation neared and the feeling of "belonging to the club" or "club mentality" invaded our minds. We were fully aware that other Marines had survived Parris Island and the same treatment we had received, and were better men for it. We had taken all the punishment Parris Island and our Drill Instructors could dish out and we had earned our Eagle, Globe, and Anchor emblems that adorned our collars and "overseas covers." This was no time to be invective and stir up a hornet's nest about our treatment. One thing was for sure. We would proudly carry the title of U.S. Marine with us for the rest of our lives. There really is no such thing as a former Marine. There are just Marines, period.

Thursday evening our Senior Drill Instructor, S/Sgt. Lawson, called a school circle and for the first time talked to us recruits in a normal tone of voice. There were no shouts this evening. We almost didn't know how to react because of the change.

After a few words of congratulations in successfully surviving boot camp, S/Sgt. Lawson asked, "Are there any questions?" One recruit raised his hand and asked, "Sir, how come the Drill Instructors treat the recruits so bad? Why are there so many slaps, punches, and kicks?"

We were all kind of shocked at the question but S/Sgt. Lawson never flinched as he replied, " When we get a group of recruits here on the Island for training we don't have time to separate the good guys from the ones that think they are bad asses. We need to find out in the first few days who will take orders and who will try to be a complete knucklehead. When we weed the troublemakers out we can begin to train those that will comply with our orders. This is the way we have always trained Marines and we have trained some mighty fine Marines here!"

After a few moments hesitation he continued, "When I came here for boot camp in 1952 I arrived here by plane. After our plane landed at Page Field the Drill Instructors came up the ramp and into the plane to escort the recruits off. By the time I got down to the tarmac I had had my ass kicked three times! You recruits have had it easy compared to what we faced in the "Old Corps." You better be glad we've eased up some since then!"

When he said those words I'm sure ninety six recruits were thinking along the same lines. If the "attitude adjustments" were worse back then they sure must have been very bad indeed! We didn't think it could get too much worse than we had experienced.

Friday morning we had reveille at 0400 hours to prepare our summer uniforms for graduation. At 0500 we went to chow then went back to the barracks to change into those uniforms. We then had an inspection in the street outside our barracks which was conducted by all four of our Drill Instructors, S/Sgt. Lawson, Sgt. Shue, Sgt. Garcia, and Corporal Meade.

They measured to make sure the distance between our Shooting Medals and National Defense Ribbon was proper in relation to the top of our uniform pockets. They also made a last minute check for "pennants" (strings) which might be showing on our uniforms.

When they were satisfied with our appearance we began our march to the Parade Deck (better known as the Grinder) located near the 2nd Battalion area. This would be the last time we would have to make this march as a platoon.

Arriving near the reviewing stand we could see that the bleacher seats were already filled with civilians and military personnel. Our series of three platoons lined up by platoon number. Platoons 380, 381, and 382 and passed in review in front of the large crowd of people. The day was already hot but for once we barely noticed the heat as the band began to play various martial airs by the former Marine Major and Band Leader John Philip Sousa.

The platoons halted in front of the reviewing stand and faced the crowd as the Commanding General gave a speech and the Chaplain said a prayer. The awards ceremony followed in which the Honor Graduate of each platoon was presented a Certificate and the best Marksman award for each platoon was also presented. Private Stevens received the Honor Graduate and Private Habermann received the Honor Marksman Award for our platoon.

The National Anthem was then played, followed by the Marine Hymn. Both tunes made chills run up my spine that day and I'm sure I wasn't the only one that felt that way while standing at attention or seated in the bleachers.

S/Sgt. Lawson then stepped to the front of our platoon and shouted," Marines of platoon 381, dismissed!" This was the first time we had ever been addressed as Marines by any of our Drill Instructors or anyone else since arriving at Parris Island. At that moment we all felt we had finally earned that title and that all the pain and agony we had endured to earn it was well worth it.

I ran towards the reviewing stand and saw Wanda and the family for the first time that day. We hugged and there were smiles all around as we weaved our way through the crowd on the way to Dad's car. I had some news for them that I had received only that morning after chow.

S/Sgt. Lawson informed me that the leave I had requested had been approved and that I could travel home with the family if I wished to do so! I informed the family that unless they wanted to stay longer I could leave immediately. They were ready to go, so Dad drove me to the 3rd Battalion area to pack my seabag for the trip home. Several other recruits were also packing their gear to go home for a ten day leave as I hurriedly gathered my uniforms and extra boots and placed them inside my seabag.

After getting everything together I shook hands with several friends and wished them good luck as we parted ways. As I was going down the ladder I met S/Sgt. Lawson coming up. He stopped and shook hands with me and again congratulated me on my successful completion of Boot Camp. His last words to me were, "Good luck Private Nickels! Maybe we'll meet again in the Fleet someday."

Dad was parked just outside the barracks so I didn't have very far to carry my gear. He had earlier met some of my Drill Instructors and said to me, "Your Drill Instructors seem like pretty good fellows!" I looked at him and grinned as I said, "Yeah, they're ok I guess."

Going back along the Causeway that led from the Island and over Archer's Creek brought back the memory of the night I had first traveled that lonesome road. I felt much better going back across that waterway than I had the night I arrived here, that's for sure.

We were all talking a mile a minute as we traveled towards home on U.S. 95 which we would be our route of travel for a few miles until we reached U.S. 26. I remember telling Dad to stop at a grocery store as soon as he could so I could buy a pop and a bag of potato chips. Doing without my favorite snack for so long had me nearly suffering withdrawal symptoms.

It seemed as though those ten days flew by. I visited with friends and family nearly every day. Some other friends visited me at Dad's house. Wanda was still staying there for now but I hoped to be able to rent a place after I learned where I would be permanently stationed. Of course that would depend on whether or not I was sent straight to Vietnam after ITR.

My orders called for me to report back to Parris Island to wait there until my orders to report to ITR were issued. I dreaded leaving Wanda again but she was resigned to having me leave home for at least another few months.

Dad offered to take me back to Parris Island and I really appreciated him giving me a ride back. The only money I had received so far from Uncle Sam was $26.00. Wanda had by now received a small check also. I doubt that Wanda and I had enough money between both of us to pay for a bus ticket back. Besides, I didn't need any money once I made it back on the base. She would need the few dollars she had until I could get my regular $44.00 paycheck.

Dad and I left home Sunday, May 29, 1966 to travel back to Parris Island. The parting with Wanda was painful as I hated to leave her not knowing when, where, or if we would ever be together again. I still had four more months of training left before I would be assigned to a permanent duty station and I knew Wanda didn't want to be a burden on my Dad's family by having to stay with them. I had much to brood about as we started on our way to South Carolina that early Sunday morning.

It was already dark as Dad pulled into the 3rd Battalion parking area. I thanked him once again for driving me down and cautioned him to be careful as he returned home. I don't remember for sure but I think he drove straight back home without getting any sleep until arriving home the next day.

I checked into the Headquarters Building where the Duty NCO read my orders and directed me to the Casual Platoon area where I would wait a few days for my orders to Camp Geiger. The Casual Platoon Barracks were attached to the Headquarters Building so I didn't have far to travel to find an empty rack. It was after 2200 hours so the other Casual Platoon members were already in the rack. I tried to keep the noise down as I settled in to sleep. I went to sleep wandering what the Marine Corps had in store for me in the coming months. I was looking forward to getting my training over with and settling into some kind of routine. It was the unknown factors that worried me the most at this point.

Platoon 381, 'S' Company, 3rd Recruit Battalion, -Parris Island, South Carolina May 20, 1966-U.S. Marine Corps Photo

Ronald Allen
Thomas Allen
R. A. Andreoli
James Bailey
D. T. Balfour

R. R. Berardino
Terry A. Bishop
S. M. Breneisen
Hans R. Bronner
Conrad Brown

Donald Bruski
Maurice Burton
George Butler
Barton W. Butts
J. R. Carrington, Jr.

R. C. Carver
James F. Charls
M. J. Christmas
Aaron Couper, Jr.
Robert S. Corwin

Richard R. Cote
Gary R. Crawn
Joseph W. Davis
Ralph G. Denena
Gene B. Dixon, Jr.

Dennis Domagala
Paul E. Doty
Donald W. Drake
Ronald Duva
John R. Faraone

Darrell Fleming
M. P. Frederick
R. L. Gambarale
Glenn A. Graf
John Habermann

Donn Heitmann
Carroll G. Hill
Edward Hoffman
Norvell Jones
John E. Keating

James M. Kinder
Ricky J. Labbe
Robert Lafleur
Thomas Laneave
Albert Lautier

Edward G. Leite
Richard A. Levy
Richard A. Lewis
Willie Lovette
James E. Lux

Edward Maddock
P. M. Marciano
James McArthur
Michael McCaw
Robert L. McRae

Francis Miller
J. N. Miller
Robert A. Mosher
Henry Mumme
Eddie G. Nickles

Joseph P. Owens
Jeffrey A. Peck
Freddie A. Roach
James Robertson
Edward J. Ross

R. S. Schrader
S. W. Shifflett
F. J. Sledjeski
Thomas Staples
Raymond Stevens

P. G. Strosnider
C. W. Sunley
Alex J. Terlesky
John Theriault
William Todd

James D. Tussing
T. D. Walton
Jerald Weaver
Nathan R. Weiss
T. R. Wekenmann

John L. Welsh
Myron L. Velverton
R. M. Englehardt
D. H. Dale
D. J. Giancarlo

D. Arnone
C. K. Carey
N. Kantargi
P. M. Carmody
R. W. Hawkins

W. V. Hartness
R. D. Fraze
R. H. Jones
D. L. Lewis

Wanda and me March, 1966,

Camp Lejeune, Sept., 1967

Eddie Nickels

Wanda Nickels

Our son Jeffrey, my Wife Wanda, our son Steven

Our Daughter Alisha Nickels Bowling

THIRD RECRUIT BATTALION PLATOON 381 M.C.R.D. PARRIS ISLAND, S.C.
CPL. M.W. HENRY SGT. M. GAITAN SSGT. T.T. LIDSTER SGT. J.C. TODD
GRADUATED 20 - MAY - 1966

Daughter-in-law
Michelle, Son
Jeff, Grandson
Dylan.

Daughter Alisha,
Grandchildren-
Summer, Jordan,
Carly, and Son-in
Law Ricky.

Back-Eddie,Steve,Alisha,Jeff. Front-Jordan,Summer,Wanda,Kimmy,Tyler,Ryan

Grandsons

Dylan, Daniel

Son Steven,

G-Daughter

Kimberly

Grandsons,

Tyler,Ryan

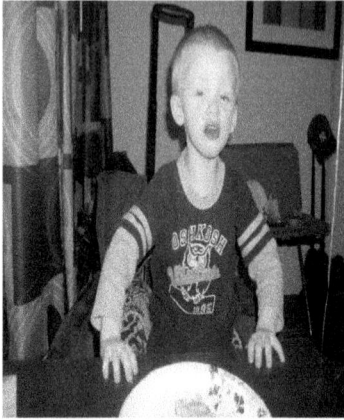

Tayvin Nickels

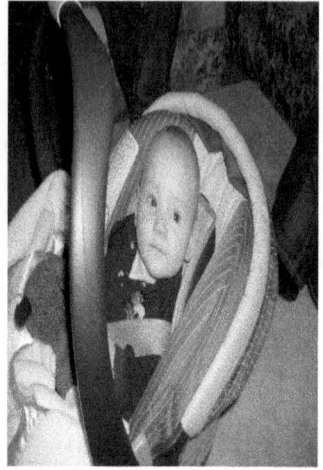

Carson Lucas

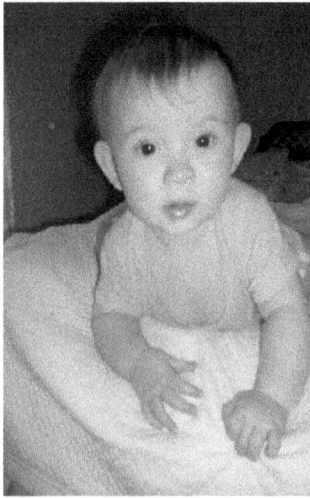

Brooklyn Nickels

McKenna Nickels

Chapter Seventeen

CASUAL PLATOON CASUALTY

Each of the four Recruit Battalions that made up Parris Island's Recruit Training Regiment had a Casual Platoon in their organization. These platoons had several categories of recruits assigned to them at any given time. Some platoon assignees had already graduated and were awaiting transfer to Infantry Training Regiment at Camp Geiger, North Carolina. Others were awaiting discharge for various reasons and some were healing from injuries during training and would resume training upon recovery from these injuries. Like myself several already graduated Marines were returning from leave and were waiting to join a newly graduated series of platoons as they boarded the buses to Camp Geiger. Usually a series of platoons left for Camp Geiger every Saturday and I expected to leave with them in a week or so.

My first morning in Casual Platoon started with a 0500 reveille call. As I listened to the familiar notes of the bugle sounding reveille throughout the 3rd Battalion area I quickly made my rack up and glanced around the small squadbay to familiarize myself with my surroundings. I saw several recruits with casts on their arms, legs, and feet, and most were having some difficulty struggling to stand as they tightened their sheets and blankets on their racks.

There were about thirty young men in the platoon that morning and about half of those appeared to have casts on their limbs. The other half, including me, were ready to proceed to ITR as soon as another Series of Platoons left the island. As for me, I was anxious to get off the Island and resume my training as soon as possible.

The NCO in charge of our platoon, a sergeant, held a formation inside the barracks and then marched the able bodied Marines to chow. Those on crutches and in casts were allowed to proceed at their own pace to the mess hall. As we marched from the Headquarters building to the nearby mess hall I could hear the familiar shouts and cadences of the 3rd battalion Drill Instructors as they marched Recruit Platoons on the Parade Deck and headed to chow. I was pleased that I wouldn't have to endure that part of training anymore.

Several platoons of recruits were lined up and waiting to enter the mess hall and I heard the familiar shouts of "Step, face, uncover, move!" as we waited to enter the mess hall. Most of those of us in Casual Platoon had already completed our boot camp training and were allowed to eat our meals at our leisure and as much food as we desired. As I ate in the mess hall that first morning I couldn't help glancing around to see if a Drill Instructor was going to approach our table and jump us for "eyeballing" them in some fashion. This relaxed mood while at chow would take some time to get used to after months of sheer terror while eating chow as a member of my training platoon.

After chow we marched back to the Headquarters where the daily routine was explained to me by the Sgt. In charge. While I was in Casual I and the other platoon members would be required to clean the squadbay area immediately after morning chow. I would then have to work at an assigned job until 1600 hours. Evening chow was at 1730 hours and after chow all platoon members had to field day the complete 3rd Battalion Headquarters Building until 2100 hours. It was obvious that our free time would be limited while in the platoon. We would however have Sundays free to do as we liked, with full PX and base privileges. While he was explaining all that, I was thinking that with any luck I would be out of here before next Sunday and wouldn't have much cleaning to do.

Some of the more able platoon members were assigned to daily working details but because of my Basic Administration MOS I was assigned to the Unit Diary Section of Headquarters which was located just a few feet from our Casual Platoon Barracks area. Our Sgt. led me to the office and introduced me to the L/Cpl. who would be my boss while working there. I don't remember the name of the L/Cpl. But I do remember the song that he hummed over and over all day long as he and I worked in the Unit Diary section over the next few weeks. *"My Baby Does the Hanky Panky"* would reverberate throughout our work area nearly every moment of the day until 1600 hours. I finally asked him to change songs for awhile at least, but to no avail. He said that song was his favorite and that "I could no more drop that song than I could drop my girl friend," as he put it. I will forever remember that song and the L/Cpl. that sang it nearly all day long, every day, as we worked.

The room we worked in was very large and contained several desks where other Marines, permanent personnel, were busily doing the paperwork required to run a Recruit Battalion at Parris Island. The Administrative Chief, an E7 Gunnery Sergeant, had a desk near our Unit Diary section and would occasionally require the L/Cpl. and me to work late into the night to catch up on our work.

We had the responsibility of using metal cards of about 3"x 5" dimensions to emboss the pertinent information of each Recruit in training in the 3rd Battalion. This information included the Service Number, platoon number, training dates, training status, and assigned MOS of each recruit as assigned at Headquarters Marine Corps. This required that we handle literality thousands of pieces of information to be embossed on metal cards each week. Working overtime well into the night was the norm during the few weeks I was assigned to Casual Platoon.

We would receive several manifests of information for a series of platoons in our Battalion that contained the information we needed to transfer to the metal cards. I always enjoyed scanning over the different MOSs assigned to the recruits. We always saw them first and if we had been so inclined, we could easily have changed each recruit's MOS to one other than had been assigned to them from Headquarters. We had control of what was put on their official cards and our work wasn't questioned. I would read a Recruit's name beside a 0300 (Infantry) MOS while thinking how easy it would be for me to change him into a truck driver or an artilleryman, but of course I never did. I had been trained to follow orders and that's what I did.

The L/Cpl. I worked with liked to party nearly every night in nearby Savannah, Georgia. After the Chief Administrator, a Gunnery Sergeant, left the Headquarters Building, the L/Cpl. would caution me and the other Private that sometimes helped in the Unit Diary to "Make sure you finish embossing the Recruit data before you secure (quit) tonight." With those words he would slip out of the building and rush off to Savannah, leaving us to work late into the night to finish the job.

The Gunny finally caught on to what he was doing and called me into his office one evening to ask if the L/Cpl. was leaving me by myself every night to work while he partied in the bars of Savannah. The Gunny had obviously grown tired of what he was doing each night. Not wanting to rat him out I denied that he left me to finish up by myself most evenings. The Gunny cautioned me to not lie for my boss and although skeptical of my answer, dismissed me to go back to work.

A few nights later, after everyone had left the office for the evening, the L/Cpl. again left me and the other Private to complete the posting of orders for the Recruits as he left for Savannah. He picked his cover off his desk and turning to face us said, "You guys finish up here before you leave for the night. If anyone asks, tell them I got sick and went home!"He went out the door singing his theme song, "My Baby Does the Hanky Panky." I had never mentioned to him about the Gunny's questioning of me about his skipping out early every night but he was well aware of the danger of leaving his post without permission. The Gunny had actually called him down for it in my presence a couple of times already.

A short time after the L/Cpl. left, I looked up from my desk to see the Gunny walking into the office. He walked to my desk where I and my helper was busily transferring orders from the manifest to the metal plates and asked, " Where is your boss?" I stood up and answered, "Sir, I have no idea where he is, Sir. He left a little while ago, saying that he was sick and going home." Without replying he headed for his desk and opened a drawer and pulled some papers from the drawer. He walked back towards myself and the other Private and handed each of us a blank sheet of paper as he said with anger in his voice, " I want both of you to write that your superior left you to do his job and yours too while he left for Savannah. I looked at him and said, "Sir, he told us he was sick when he left."

Glaring at me he said, "Obviously both of you have been covering for him while he was supposed to be here working. If you refuse to write him up I'll see that neither of you are on the buses to Camp Geiger when they pull out of here! You'll stay here for a year if necessary until you make a statement about him! You two don't owe him anything. He's been putting extra work on you so he can paint the town. Now get busy writing those statements!"

My buddy and I looked sheepishly at each other as we sat down at our desks to write our statements. We both knew that the Gunny was perfectly capable of putting a legal hold on us if he chose. We had no choice but to comply with his order to write the statements. To not do so was to risk an obstruction of justice charge against us. Although neither I nor the other Private had done anything wrong I felt like I was betraying the L/Cpl. who not only was my boss but who had become a friend to me the few days I had been working for him.

241

I never again saw the L/Cpl. that we were forced to write up that evening. I have hoped all these years that his punishment turned out to be very lenient for him. I sometimes hear his favorite song played over the radio which never fails to bring back memories and regrets of those days when I worked in the 3rd Battalion Unit Diary and was forced to write up my boss.

On evenings when I didn't have to work late in my Unit Diary job I had to help Field Day the whole 3rd Battalion Headquarters building. I had become friends with a Private from New York and a Private from West Virginia who were, like me, waiting to ship out with the next series of recruits to Camp Geiger for ITR.

Private Donald Millner from West Virginia was the almost exact likeness of the popular country music singer, George Jones. The first time I saw him in the barracks area I actually though that he might be George Jones himself. He could have easily passed for his twin brother.

The other friend, whose name I have forgotten, was a Catholic who had at one time been studying for the Priesthood. He and I had many discussions about the differences of our religions, with none of them ever leading to harsh words or anger of any kind. I kidded with him a lot about his being a "religious beer drinker." He got me back about being a "religious teetotaler." He and I had the job of using the electric buffer to shine the decks of the building which took nearly four hours each evening to complete. (I helped him only if I wasn't working late in the office.)

Each of us had some laughs at the expense of the other when we first tried to use the buffer but we both finally mastered the technique and no one challenged us for the job. I would run the buffer for a couple of hours while he handled the power cord, then we would switch jobs for the next two hours.

There was little stress on us while in casual Platoon. Those NCOs in charge of us would assign our jobs in the mornings then would not come around again until lights out in the evenings. I would have been contented to have spent my whole time in the Corps working and living in the 3rd Battalion Headquarters.

While our platoon was performing a Field Day cleaning of the Headquarters building on Thursday, June 2nd, 1966, I was bending down alongside a baseboard of one of the bulkheads and cleaning the baseboard with a wet cloth when I heard and felt a bone crack as it broke on the top of my left foot. I called the Corporal that was supervising our detail to report the possible breaking of my foot and told him I was in quite a bit of pain from it.

He ordered one of the other Marines to assist me to the West End Dispensary to have the injury checked out. I walked, or rather hopped to the dispensary the best I could but the pain got worse with each step. My Marine buddy supported me by holding me up with my arm across his shoulder to take some pressure off my foot while we walked to the dispensary.

After the Navy Corpsman took some x-rays of my foot he told me that I had a stress fracture on the top of my left foot and that he was sending me to the Main Medical Dispensary located on the Main Base to have a Doctor check the fracture. I would have to wait until tomorrow to see the Doctor because of the late hour and because a stress fracture wasn't considered an emergency.

The Corpsman gave me a pair of crutches to help me walk back to the barracks at 3rd Battalion. I was in severe pain when I arrived back at the barracks that night.

The next morning I again hopped all the way to the Main Dispensary with the help of my pair of crutches. I never thought of asking for a ride there. I was used to doing exactly as I was told without complaint so I walked all the way to the Main Base Dispensary with my fractured foot. I had to walk alone this time.

Arriving there pretty much exhausted by my trip, I was examined by a Navy Doctor and had another x-ray. He confirmed the diagnosis of the Corpsman. I had a stress fracture of the top of my left foot. He said I was going to be put on a Medical Hold because of my injury and would be unable to proceed to Camp Geiger as scheduled this weekend. The Medical Hold would be effective until my foot healed which would take approximately four weeks.

I felt devastated with the news of my Medical Hold which would keep me on Parris Island for at least another four weeks. I was looking forward to completing my training as soon as possible so that I might be able to move my wife to my permanent duty station. This set back in my plans was hard to take after so many other disappointments I had suffered along the way. I had no choice but to obey the Doctor's orders and wait until my foot healed.

The Doctor said it wouldn't be necessary to put a cast on the foot. He bound it tightly with an elastic stocking or bandage type dressing and cautioned me to stay off the foot as much as possible. I had been thinking more along the line of a nice hospital bed to recuperate in where I wouldn't have to run the old deck buffer every night or work until 2200 hours in the Unit Diary Section every night. Instead I would now have to do the same job while hopping around on my crutches. My friend, the Catholic Priest-to –be would really be able to laugh at me as I stood on one foot and buffed the decks each night!

The fact that my work area was situated so close to my sleeping area was a lucky break for me.(No pun intended.)I could easily walk the fifty of so feet from my rack to my desk in the Unit Diary Section of the Headquarters Unit without much pain or effort. The fracture I had suffered didn't alter my daily routine at all because of the close proximity of my Barracks to my work area. Going to chow required quiet an effort and much time to complete due to my injury. I avoided going there as much as possible because of the fracture.

I still spent my days organizing the print –outs we received each day that had to be transferred to metal 3"x5" cards. I checked my own embossed card occasionally just to make sure that someone hadn't somehow changed my MOS to an Infantryman instead of a Basic Administration Man MOS that I had been assigned during boot camp. That would have been a cruel joke to have been played on me, considering the trouble I already had with the pain and aggravation that my stress fracture had caused me. I hadn't figured on becoming a casualty while I was assigned to Casual Platoon! It was somewhat ironic that I had done so after surviving the rough physical training of boot camp.

Several days after my foot injury I received word by Guard Mail that I was to report to the Main Dispensary the next day for further evaluation of my foot fracture. I was confused by this summons as I had another appointment for next week to monitor my healing process.

I hopped to the Dispensary the next morning and reported to the front desk. I was directed to a room where several Marines were seated around the room which was devoid of any other furniture. Some Marines seated there had casts on their feet and some didn't. It was obvious to me that this evaluation had something to do with foot injuries all of the seated Marines, including me, had suffered.

Soon, several white coated doctors entered the room with each of them holding a clipboard with some paperwork attached to them. After some quiet discussion one of the doctors ordered all of us to stand up and remove all our clothing, including our skivvies. The dozen or so of us looked at each other with questioning glances as to what was going on. Not in the habit of disobeying superior officers, we slowly began to remove our clothing, including our shoes and socks as far as our sore, fractured feet would allow. I remember that my foot was still swollen because of the fracture and the removal of my sock (which was under the elastic binder) was fairly painful for me.

After undressing for the doctors, (assuming they were doctors), we stood there completely exposed and wandering what this event was all about. The assembled group of observers that were making notes on their clipboards was exchanging comments with each other while they stood about 25 feet from us, like we had the plague or something. Two Navy

Corpsmen were standing to the side and seemed to be there to give us our orders.

Several photographers were taking pictures of our naked bodies as the clipboard holders continued taking notes. After a few minutes the flashbulbs stopped flashing and the Navy Corpsmen ordered us to get dressed. The assembled group of doctors and photographers left the room with much discussion taking place among them as they left.

When the last white coat went through the hatch he shut the hatch behind him, leaving us to wonder what that incident was all about. One of our group of Marines asked the Corpsmen, "What in the hell was that all about?" One of the Corpsmen replied, "Those were a group of physicians that are writing a medical report to be included in a medical book about why so many Parris Island Recruits are suffering stress fractures during training. They have noticed that the fractures have occurred mostly among tall, slim recruits and they are curious as to why that is happening."

As I looked around I noticed that all Marines in our group were indeed tall and slim. We were all also victims of a stress fracture to one degree or another. I wasn't too pleased to learn that our photos were to be included in a college medical book and circulated around campuses across the nation!

One of our group of Marines spoke his mind and said to the Corpsmen, "It would have been nice if we had been asked if we wanted to be in a medical book somewhere. Evidently we don't have any rights left when we arrive here!" One of the Corpsmen looked at our group as he opened the hatch to let us leave and replied, " Yours is not to reason why, yours is to either do or die!" That was the first time I had heard that comment but it wouldn't be the last as I would hear it many more times during my military service.

I was sitting at my desk and looking through a new manifest of orders we had just received when I saw my own name on the list to report to Marine Corps Basic Administrative School at Parris Island on August 1, 1966. I knew that I would never make that class on time unless I left for Camp Geiger by the end of the current week. I received those orders June 9, and I was aware that a series of platoons were leaving our battalion for Camp Geiger June the 11th. If I could leave with them I just might be able to make the school in time after going through ITR. If I missed this class my MOS would be changed for sure and I didn't want to be reclassified as a 0300 "Grunt" if at all possible.

Since I was on a medical hold I would have to try to convince my Navy Doctor that I was fit as a fiddle for training. I had no choice but to hop over to the Main Dispensary to see if I could see the Doctor even though I didn't have an appointment.

I took off after noon chow and hopped almost two miles to the dispensary, in pain all the way. The pain didn't bother me too much since I was on a quest to try to achieve my freedom from this danged Island. I felt that I was destined to remain on Parris Island the rest of my life with no hope of ever leaving. I wanted to be with those graduating recruits when they boarded those buses this weekend in the worst way. The Navy Doctor, a Lt. Commander, held my future in his hands and had the authority to set me free if I could somehow convince him of my miraculous healing after just seven days of recuperation.

After a long, slow walk I arrived at the Dispensary and was making my way down the passageway to his office when to my surprise I saw him and a couple of other doctors come out of his office and head in my direction. I assumed the position of attention against the bulkhead as we were required to do when an officer passed an enlisted man in close quarters. As the trio of doctors passed me I blurted out, Sirs, Private Nickels requests permission to speak to the officers, Sirs!" I was so flustered that I couldn't remember the rank of my Doctor, nor his name. They stopped in front of me and my Doctor spoke up, "Which Doctor was you needing to see, Private?" I replied, "Sir, it's you, the Lt. Commander I need to see, Sir!" He said, "At ease Private. What can I do for you today?"

I quickly explained my important mission to him about how much I wanted to leave this very weekend to continue my training. I assured him that my foot was completely healed after only seven days and I could easily make every ten mile run required of me in ITR. I just needed for him to release me from medical hold and I would just run back over to 3rd Battalion and begin squaring my gear away to prepare to leave this weekend!

He looked me in the eye and said to me politely but firmly, "Private, your enthusiasm to continue your training is commendable but I'm afraid I'm going to have to continue your medical hold for at least a couple more weeks."After hearing those words I had a sinking feeling in my chest but I again assured him that I felt well enough to begin my training that very day if necessary.

As the three men began to walk off my Doctor ended the conversation by saying, "When you return for your next appointment we'll see how well you're doing and go from there."

My efforts had been in vain and I had suffered pain and misery in my foot walking over here for nothing. I had no choice but to abide by the Doctor's decision. I headed back down the sidewalk with a heavy heart, an aching foot, and a bruised ego. My powers of persuasion needed some work, that's for sure.

Exactly two weeks later I received my new orders which also included a change of MOS as I had expected. My MOS was changed from 0100, Basic Administration Man, to 3000, Basic Supply Administration and Operations Man. I didn't feel too bad about the change because I felt a supply job wasn't that much different from my first MOS. My school date for Supply School read "no later than August 10, 1966."

I again hopped over to the Main Dispensary to seek the permission of my Doctor to leave the Island. The next series of platoons from 3rd Battalion were leaving Monday, June 27, and I intended to be on one of those buses even if I had to stow away this time.

I managed to see the Doctor in his office this time. I explained to him my situation with the receipt of my latest orders and made my best argument why I was now ready to ship out of here. Without saying anything he examined my foot and had me walk across the room and jump up and down a few times. These actions hurt like the devil but I managed to hide my pain and misery as I performed for him. He sat down at a small table and filled out a "Release from Medical Hold" form to take to the Lieutenant in charge of Casual Platoon. As he handed me the document he wished me good luck with my ITR training and cautioned me to "take it easy with your foot until it completely heals." I didn't have any idea how one could "take it easy" while undergoing combat training at ITR but I didn't argue with him.

I left his office in a very good mood and hardly noticed the pain in my foot as I limped and bunny hopped back to the 3rd Battalion area. I was ready to move on to Camp Geiger. That evening I dug my little New Testament Bible from my foot locker and read several verses about faith. I was needing assurance that my injury would heal quickly.

Chapter Eighteen

"WHIP SNAPPERS"

By 0600 hours on Monday morning several Trailway buses began arriving at the 3rd Battalion Parade Deck to transport the recently graduated Recruit Series to Camp Geiger, North Carolina. I had now been on Parris Island for 13 and 1/2 weeks and I was more than ready to move on. I had my seabag, clothing bag, and small duffel bags packed and ready to go with them. I had even skipped chow to make sure that I would be on one of those beautiful buses that would finally carry me away from there. I was a little afraid that somehow my Navy Doctor would change his mind and block me from leaving that morning. If he had known that my foot was far from being well, he would have surely stopped me.

When I carried my gear outside the Headquarters Building the extra weight caused my foot to throb with pain even more intently. The pain didn't matter as long as I was on that bus. I would deal with the consequences later. I was just hoping I would be completely healed by the time we started ITR.

Several of my Casual Platoon buddies were also leaving the Island with me to complete our Infantry training at Camp Geiger. Private Millner, the George Jones look-a-like, was one of those buddies that would be leaving with us. He had been assigned to a different bus, but we would later be assigned to the same company at Camp Geiger.

If I remember correctly there were six buses in our convoy that left Parris Island that morning. We traveled on the now familiar causeway that traversed over a swampy area, then crossed Archer's Creek and passed the kiosk where the white gloved Marine standing there saluted each bus as it passed by.

The buses turned left at the red lights just beyond the Main Gate and proceeded to Route 17 which we would take all the way to Camp Geiger.

As we traveled north I felt relieved that I was finally leaving Parris Island to continue my training. It seemed like I had been on the Island nearly fourteen months instead of the nearly fourteen weeks I had spent there. I still had three months of training remaining before I would be assigned to permanent duty at some yet unknown Marine Corps Facility.

It was nearly two hundred miles from Parris Island to the gate of Camp Geiger in North Carolina. We left the Island at 0800 hours and passed through the city of Charleston about one hour later.

As we crossed the bridge in the City that crosses to Mount Pleasant I was in awe of the many Navy ships that we saw moored in the Cooper River. The bus driver remarked to us that those ships were part of the Naval Fleet Reserve Force. There seemed to be everything from merchant ships to destroyers that were moored there. My first view of the City of Charleston and the many ships docked there was mighty impressive for someone like myself who had grown up reading so much about historic Charleston, but had never had the opportunity to see it before this trip.

About two hours later we entered North Carolina and passed near the majestic 35,000 ton Battleship *USS North Carolina* moored in the Cape Fear River. The Battleship (BB55) had been moored near the City of Wilmington, North Carolina since October 2, 1961 as a permanent Historical Memorial to this great ship.

Its armament included nine 16 inch guns that fired shells weighing nearly two tons each! Other weapons on this ship were twenty turret mounted 5 inch guns and many, many 20mm and .50 caliber antiaircraft guns mounted along both sides of her decks. The USS North Carolina was a formidable ship that played a huge role in winning World War Two. I had read much about her and was thrilled to see her, even if my view was only a quick glance from our bus window.

A little over four hours after leaving Parris Island our bus turned off at the entrance to the New River Air Facility near Jacksonville, North Carolina. This was also the entrance to Camp Geiger where the home of Infantry Training Regiment was (and is) located.

Camp Geiger is part of the sprawling Camp Lejeune Marine Corps Base which is located a few miles further up the road on route 17. This would be my home for the next five weeks or so.

As we traveled past the buildings that made up Camp Geiger I realized that this facility must have been in use for many years. Almost all the buildings looked as if they had been in use since at least World War Two and maybe even longer. I wasn't overly impressed with my first view of our new home. Most of the structures were constructed of wood, with very few brick buildings in view.

Our buses proceeded to a large open parade deck where we filed off the buses, reclaimed our gear which we had stored in the storage compartments, and got into company and battalion formation. As we stood there at attention I saw several Marines from our group of buses who were standing at attention with NCOs busily screaming at them and giving them some punches to get their attention.

Shades of Parris Island! I hadn't seen any such conduct since I graduated from recruit training. I began to think I was mistaken in thinking that the" rough stuff" ended when one completed their "Boot" training. I don't know the reason for the rough treatment that day but I think the NCOs wanted to put the scare in us right off the bat so as to get our attention. All three hundred of us new arrivals realized at that moment that some things never change when you're still in training. I was feeling a little down to realize that we would get little respect until we accumulated a little rank further on down the line.

We marched to chow and got into a very long line to await our turn to enter the mess hall. The cooking odors wafting from the mess hall made me feel very sick that day for some unexplained reason. It might have been a combination of the trip from the Island, the steaming hot sun, or just generally being in a place where I would prefer not to be. I was missing Parris Island at that point.

As I got closer to the building I saw where someone had scrawled "Kilroy was here!" and other graffiti on the side of the building. Nearly every one familiar with World War Two lore knows that American troops scratched that phrase on buildings throughout Europe as they advanced from town to town and battle to battle. For some reason those words caused a rush of sickness and homesickness to pervade my body at that moment. If I hadn't already been so close to entering the mess hall I would have fell out of line and found a shade somewhere to recuperate.

I finally reached the metal trays, forks, and spoons and picked up one of each. The soap stains on those utensils and on the tray made it look as if they had been washed but not rinsed. I had seen mess utensils in this shape once before when I had my first meal at Parris Island, but never again until now. I was afraid to chance putting any food on top of the soap stains. I merely picked up a glass and poured some orange juice from the dispensing machine. My first meal at Camp Geiger wasn't very promising for my future culinary experience.

That first meal at Camp Geiger turned to be the only bad experience I had with eating at that facility. Every mess hall I had the privilege to eat in after that first day had excellent food and outstanding cleanliness. I was afterwards glad to see that my first impression had been wrong.

After chow we were all marched to a cluster of buildings with a large dusty field fronting them. We were carrying our seabags, ditty bags and everything else we had left Parris Island with. We were pretty much weighed down with our gear but we had still more items to add to our heavy loads. We were in front of a supply building to receive an issue of 782 gear and our rifles which we would be using while at Camp Geiger.

We were in company formation facing the dusty dirt road which wound through the area of the supply buildings and waiting for our new gear issue when a large company of deeply tanned Marines came marching up the road. As they came abreast of where we were standing they were halted and came to "Order Arms" with their M1 rifles they were carrying. I could hardly believe my eyes when I saw the M1 Garands. I didn't know the Marines still used those older weapons in any capacity anymore. I wondered if we were going to be issued the same rifles as these dusty Infantry Marines were carrying.

This new company of Marines was there to turn their equipment in to supply. One squad of them formed a line while the other squads stood in formation. It was then that I heard a voice calling in a loud whisper," Nick! Nick! Hey Nick!" Since Nick was the name my friends had called me while in boot camp I directed my attention to the Infantrymen lined up across the road from us.

I wasn't sure if the voice was calling me or someone else but as I scanned the group of Marines I saw a hand holding a rifle at Order Arms while the other hand was waving at someone. I looked up to the face of that Marine and instantly recognized

Private Wilson, whom I had last seen on our May 20 graduation day in boot camp. He had been one of the graduates who had proceeded directly to Camp Geiger from Parris Island upon graduation while others, including me, had received a ten day leave.

I recalled that he and my other friend, Private Moore, had received an infantryman MOS at the end of boot camp. While I had been stranded in Casual Platoon with a stress fracture and while having to submit to having naked photos taken of myself by a group of doctors, they had been in ITR and had finished their training on the very day of my arrival in ITR! I realized then that I might have been better off to have received an infantry MOS after all. It was too late for regrets now though.

Private Wilson spoke to me out of the side of his mouth and said, "Where have you been, Nick?" In low tones so as to not bring down the wrath of a NCO down on me I replied, " I've been on leave and on medical hold on Parris Island with a fractured foot. What's your status Wilson?" He again whispered, " We're on our way to Vietnam after today!" I didn't know what to say other than to reply, " I'm really sorry to hear that. I hope everything goes alright for you!" He had become a new Father while in boot camp and had hoped that he might escape going to Vietnam but it was not to be. In a couple of minutes his squad was marched to the supply area to turn in his gear. The last time I saw him he was standing at attention with his blanket in his arms while a black NCO was shoving and pushing him as if there was a disagreement of some kind between them. I have always felt that he would have been highly embarrassed if he knew I had witnessed that scene. I never saw him or my other friend, Private Moore, again.

It took several hours to issue all the equipment to the large company of men that our several busloads of Marines had now been assembled into. I was pleasantly surprised to find that we would indeed be training with the old M1 Garand of World War Two and Korea fame. Along with our rifles we were issued a green plastic canteen, a cartridge belt, a shelter half, a mess kit, a knapsack, a field pack, tent pegs, a helmet, helmet liner, gas mask, bayonet, and a first aid kit. There may have been a few other items issued also.

After a complete equipment issue we marched to chow at the unimpressive mess hall where I had endured the first meal at noon. This time I felt a little better and the food looked a little tastier than it had looked to me at noon chow.

With the evening meal under our belts a couple of NCOs marched us to a large barracks where a red sign with yellow letters identified our next home as "Headquarters and Service Company." We were divided into three platoons with each platoon being assigned to one of the three decks of the barracks. I was assigned to the second floor along with my Casual Platoon friend, Private Millner.

After stowing our gear we were assigned to cleaning details for our squad bay area. The cleaning went on for several hours and ended only after an inspection by a Staff Sgt. proclaimed the area to be sufficiently clean. This happened only after two or three repeat cleanings of the barracks. It's never good enough on the first inspection it seems.

After the cleaning inspection we prepared for a personal inspection which included fingernails, hair length, and uniform inspection. It was 2030 hours by the time we had finished the cleaning and stood all the inspections. We had been through a long day and fatigue was beginning to be a factor with all of us that had left Parris Island together earlier that day. The day wasn't over yet and another unusual incident awaited us before we were able to hit the rack for the night.

Just before 2130 hours we were ordered outside the barracks and into formation by our four NCOs and one Lance Corporal that had been overseeing our cleaning details and other activities after our assignment to H&S Company earlier in the day. At this late hour the normally busy atmosphere of activity around the area had subsided and the officers and higher ranking NCOs had left for the day. We were alone with our five "handlers" as we stood in formation in yellow footprints that were just like those we had first stood in at Parris Island.

The streetlights that surrounded the area threw off enough light to see but were dim enough that faces in the formation of Marines and among the troop handlers were almost indistinguishable. One thing that was fairly obvious to our company of newly arrived Marines was the fact that four of the five handlers in charge of us were carrying whips about four feet long in their hands. I had seen pictures and movies where some Nazi officers in World War Two were herding Jewish prisoners into box cars while wielding similar whips and I certainly didn't like the looks of our situation at that moment. I couldn't figure why NCOs would need for each of them to carry a whip but I feared that the reason might not be a good one for our formation of Marines. My fears turned out to be sustained.

As we stood in formation one of the handlers announced that, " When I call your name Mother F......s you better line up quickly in the street and you had better be movin'!" His tone of voice didn't bode well for our assembled group and was reminiscent of the tone we had heard many times on Parris Island. While we were familiar with the tone of voice being used we certainly had never seen Drill Instructors with whips in their hands. Again the handler shouted, "When you report to me you bastards better report in the correct manner or you'll wish you had! "

He then called a name from the papers which he had fastened on a clipboard. A Marine rushed out of one of the four squads we had lined up in and hustled in front of the NCO troop handler and started to report, "Sir, Priv..." That's as far as he got and the whip came down around his shoulder with a smaaakkkk! When I saw him receiving a blow from the whip I was caught by surprise even though I was half expecting it. He barely flinched with the blow and finished his reporting to the NCO as if it had never happened. I was confused by what I had just seen and I wondered if these NCOs were just rogues acting on their own or were they acting under orders? None of us were sure at the time but since it was night and one handler, the L/Cpl., wasn't carrying a whip, it became pretty clear to our more than 250 assembled Marines that these sorry excuses for Marine handlers were acting out of hate, spite, or for their own pleasure.

The next name was called and he also received a dose of rawhide from one of the handlers who by this time were also moving through the four squads and applying the lash if a Marine didn't move into the road fast enough or if his reporting attitude didn't suit one of them.

AS each Marine reported he was given a work assignment that he would be doing while waiting to begin our ITR. We were told that we would be waiting two or three weeks before starting our training. Some were being assigned to mess duty, some to base maintenance, and some were assigned to base guard duty.

The names were being called in alphabetical order so it took a while for my name to be called. I noticed that some were being spared from the whips and some got a double dose of leather. My friend Private Millner was one of those that got more than one slash of the whip. I later saw him when he removed his utility shirt in the head and as he checked his back to see how bad the marks were. (There were several stripes across his back.) Other Marines also removed their shirts in the head and checked in the mirror to assess the damage. Some were in much pain that night. Of course everyone was very angry but very few threatened to notify higher authorities. Everyone feared that to do so would result in delayed training and no one wanted that to happen. So nothing ever came of this incident to my knowledge.

When my name was called I reported in the proper manner and was told that I was being assigned to the base guard detail and that we would be moving in a day or so to a new barracks. As I turned around to get back into formation I felt the sting of a lick on my back which I wasn't surprised by. I had watched as most every Marine received at least one, and sometimes many more licks from the whips. I considered myself lucky to have received just one swat as I returned to the formation.

There was something degrading and humiliating about having whips being used on a group of men that were just trying to serve their country the best they knew how. That this punishment was meted out by four Marines was difficult to understand because of the close comradely nature usually felt among fellow Marines. The incidents of rough treatment in Parris Island weren't usually motivated by hate, meanness, or rancor. This incident at Camp Geiger was I suspect, motivated by at least one, and maybe all three factors.

It's hard to believe that the incident might not have been racially motivated as the four whip wielders were all black Marines. The only troop handler that never struck a lick and who wasn't carrying a whip was a white Marine. Racial incidents weren't unknown in all the military services in those days and I have always believed the cause of our humiliation could have been for this reason. At a later time in my career one of my fellow Marines on guard duty at the Second Marine Regiment on the Camp Lejeune Main Base related to me how he had been (allegedly)attacked while walking post around the Headquarters Company, Second Marines Armory, by a couple of black Marines. The Marine relating the incident to me had relieved me from walking that same post just a couple of hours earlier. It pains me to have to relate incidents such as these but I would emphasize the fact that these rogue incidents in no way represent the character or morals of the majority of Marines. At the same time it's important to tell it like it is and that's the way it happened.

A year or so later I heard that at least one of the Marines present that night had been brought up on charges concerning some of the Marines that he was responsible for. This news wasn't much of a surprise for hundreds of us that had the opportunity to avail ourselves of his and other's "hospitality" during our short stay at H&S Company a few months earlier. I never heard of the outcome of these particular charges and I don't know the extent of the involvement of any other Marines or if the rumor was even true.

It's inevitable that there are some bad apples in every society and every group of individuals. Boot Camp Drill Instructors are charged with weeding most of these incorrigibles out but some are invariably able to slip through the cracks in any organization, including the United States Marines. Thankfully their number is small and they eventually are found out by their superiors and punished. I hope those few rogue "Whip Snappers" of H&S Company got their punishment for their misdeeds and mistreatment of their fellow Marines in June 1966.

Chapter Nineteen

MYSTERIOUS PHONE CALL

Our company stayed the next couple of nights in the H&S Company barracks without any further incidents of abuse. It seemed that all the aggression felt towards us had been worked out that first night. After spending my second night there I was transferred along with about thirty other Marines to another barracks which was about a mile from H&S Company.

This was where we would stay for the next few weeks while on base guard duty. Discipline was a little more relaxed while we were on guard duty and our NCOs in charge seemed to be inclined to treat us recently graduated Marines as equals, which seemed to be a rare attribute among those we had come into contact with previously.

We were assigned to different guard units where we would be mostly guarding distant facilities associated with the Marine helicopter squadrons which were stationed at the nearby New

River Marine Corps Air Facility. The shifts consisted of four hours on and four hours off for three shifts during a 24 hour period. These shifts were scheduled for the next two weeks around the clock, including weekends.

On my first scheduled shift, which was on Wednesday night, I donned my helmet, helmet liner, my cartridge belt, my canteen, a magazine with only five rounds of ammunition, and the rifle I had been issued the first day of our arrival at Camp Geiger, and stood Guard Mount for inspection by one of our NCOs.

With the completion of Guard Mount eight of us were stuffed into the back of a Marine Corps Dodge pickup and driven several miles into the boondocks where there were several warehouses that contained spare parts and other equipment needed to operate a Marine Corps helicopter squadron.

Nickels in Marines

Parris Island, S. C. (FHTNC)--Ma[r]ine Private Eddie G. Nickels, son of Mr. and Mrs. James E. Nicke[ls] of Whitesburg, Ky., was graduat[ed] from Marine recruit training at th[e] Marine Corps Recruit Depot here.

During his eight weeks of intensified recruit training under veteran noncommissioned officer Drill Instructors, he learned small arm[s] marksmanship, bayonet fighting, and methods of self-protection. H[e] also received instruction in military drill, history and traditions o[f] the Marine Corps, and other academic subjects.

He will undergo four weeks of individual combat training and fo[ur] weeks of basic specialist training [i]n his military occupational field before being assigned to a permar[n]ent unit.

The warehouse complex I was assigned to consisted of four large warehouses which had the Helicopter Squadron's number and title painted in large letters on the front. The warehouses were situated close together which made walking my post a little boring but gave me the ability to observe them fairly easily as I walked my post.

I was soon able to settle into a routine of four hours on duty and four hours off without too much difficulty. I spent the first couple of hours off duty maintaining my equipment for the inspections we stood each time we went on duty. The rifle inspections were rough and one would have thought that we were getting ready for an Inspector General (IG) Inspection each day instead of going on guard duty.(Rifle inspections never got any easier during my time in the Corps, as inspections could come at any time and failing a rifle inspection is a definite no-no for a Marine.)

The following is a letter I wrote to my parents while assigned to base guard duty at Camp Geiger;

Saturday

Hello Folks!

Just afew quick lines to let everone know that I'm still kicking and doing fine. I'm still on guard duty and it's boring as heck! I'm on duty several times every 24 hour period, 4 hours on and 4 hours off. I don't get much sleep and very little time off but I'm getting used to long hours and very little sleep. I have received my orders regarding Supply School after ITR. They read this way;

Eddie G. Nickels, Second Marine Division, Supply Administration & Personnel School, MOS 3000, for Duty at Camp Lejeune, North Carolina. Report by August 10, 1966.

I have to hurry and close as I'm going on duty again soon and have to write Wanda.

Love, from your Son, Eddie. P.S. Tell little Kathy to write me a few lines.

After our pre-guard inspection was completed, the Lance Corporal that was our duty handler in the barracks for the day informed us that he would be allowing those of us that wanted to visit the area "Slopchute"[1] to accompany him there. I personally had no inclination to go with him but several of my fellow guard members were almost giddy with excitement to know that they would be allowed that privledge. To be able to have some fun while in a training mode was almost unheard of and I couldn't believe any one in charge of us would allow it.

After our four hour tour of duty the Sgt. of the Guard made the rounds and relieved each of us from guard duty. The familiar Dodge pickup truck carried us the several miles back to the barracks as the chatter amoung our guard detail went on and on about how they hoped the L/Cpl. would keep his word about allowing them to visit the Slopchute that evening. They wondered aloud if they were having a cruel (in their opinion) joke played on them.

When we arrived back at the barracks I sat down on my foot locker to begin cleaning my rifle as I usualy did before I caught a couple of hours sleep. As I did so I was surprised to see the

---

[1] Restaurant or diner where beer was sold.

L/Cpl. appear in the squadbay and announce, " All those that aren't scheduled for duty and want to go to the Slopchute fall outside into formation!" With those words about half of those Marines that weren't on duty grabbed their covers and headed excitedly out the front hatch. I was surprised to see that the L/Cpl. was as good as his word.

Four hours later I again stood inspection and was posted at my familiar cluster of warehouses to walk my post for four hours. I wondered how the Slopchute attendees were getting along and if they would be able to stand their normal duty when they returned.

At the end of my shift we were once again relieved by the Sergeant of the Guard and drove back to the barracks. As I entered the squadbay I saw the members of the company lined up in front of their racks at attention ,while our now familiar handler, the Lance Corporal, was berating them. As my group of ten just relieved guard duty Marines entered the squadbay area we were also ordered to attention by the L/Cpl..

He seemed to be so worked up by something that his now red facehad an almost evil look and the bulging veins in his neck seemed as if they might burst at any moment as he raged up and down the squadbay. He shouted, "If any one of you M..... f......s want to see how tough you are just step outoutside and we'll see who the best man is, you no good sons- of- bitches."

As we stood at attention I didn't have any idea of what was going on but I knew it probably involved the trip to the local Slopchute. The L/Cpl. was obviously reaping the benefit of too many alcoholic drinks as he raged on and on and strutted up and down the squadbay.

He was taking a big chance of calling out the Marines of our barracks as he was only about 5' 6" and might have weighed 140 lb soaking wet. Several Marines of our Company would easily go 6' 4" and were well over 225lb. They were obviously inclined to humor the slightly inebriated L/Cpl.. Besides, nobody wanted to be brought up on charges of striking a superior enlisted man. So we stood there at attention until he worked his rage out and wound down somewhat.

I don't know how long his demonstration had been going on before our arrival but he went on for ten minutes after our arrival. As he was rambling and mumbling to the top of his voice I caught one phrase that explained the whole incident. He said, " I'll teach you stupid bastards not to lie to an officer about me. It's not my fault you M..... f......s sneaked into a beer joint! Get me into trouble will you? You'll all pay for this!"

With those words he staggered from the barracks without giving the order to secure from attention. He went out the back exit and left our company of Guard Marines without a handler.

The Marines that had gone with him to the Slopchute confirmed that they had indeed been caught by the Officer of the Day while they were drinking beer and that the L/Cpl. had denied responsibility for them being there. We never saw the L/Corporal again while serving on guard duty. I imagine he had to go through "Office Hours" [2] for his actions and had probably been relieved from duty .

---

[2] Punishment by the Company Commander.

Wednesday afternoon I was posted as usual at my tier of warehouses and had the opportunity to watch as a Marine Battalion conducted assault landings in Marine CH34 Helicopters.Their large landing area was seperated from the warehouses I was guarding by a chain link fence but some of the choppers were landing not more than one hundred feet from me.

I had watched many, many simulated assaults into this same field since I had been assigned to this post several days ago but those landings had been conducted at night. This was the first time I had observed them during the day and I was thrilled and proud to watch as the infantrymen practiced their vocation. It almost made me wish that I had received an infantry MOS. (Only for a moment though.)

There were manuvers going on the whole time I was on my walking tour of duty and I wasn't the only observer. A group of Marine Officers and NCOs was also watching the action from just inside the fenced area near the gate that led to the several warehouses. Like me, they had to turn away from the closest helicopters when they landed so that loose debris wouldn't get into their eyes.

The group of officers was standing only a few yards from me and was watching as I entered the phone booth which was located on my post. I had to call in every hour to report my post secure and this was the method of reporting I had been instructed to use the phone by the Officer of the Day from the first day I had walked this post.

After reporting my post as being secure I resumed my walking around the warehouses.After I made a complete circuit around the area I saw a Marine Captain walking towards me and he seemed to be in a hurry. As he got closer I brought my rifle from "right shoulder arms" and held it at "port arms" as I challenged him by shouting "Halt!" He stopped a few feet from me and I then ordered him to idenify himself and to state his business.(It was evening and the warehouses were closed for the day.My challenge to the Captain was a routine one.) This day happened to be my nineteenth birthday. The Captain would made this birthday one I would never forget.

He gave me his name and asked in an agitated voice, "Marine what is your 7th General Order?" I quickly rattled off, "Sir! My 7th General Order is to speak to no one except in the line of duty, Sir!" He stared at me as I stood at attention and replied, "Then what in the hell are you doing entering a phone booth while on guard duty?" I began to explain, "Sir, I have been instructed to...." " Stop!He shouted. I don't want any lame excuses. Were you or were you not in that phone booth a few minutes ago?" "Yes sir,but...." Again he interjected with a shout, "Stop! I don't want to hear it! Do you understand me Private?" Knowing that he had the upper hand and that this was an argument that I could never win I decided to play along with him. "Yes sir, I understand perfectly sir!" He then said in a calmer voice, "Private don't ever let me catch you disobeying one of your General Orders again. Do you understand me?" I quickly replied with the same words I had already used, "Yes sir, I understand perfectly sir!" With those words he turned on his heel and walked back to his companions who were standing 50 yards away. I continued to walk my post although shaken by his false accusations. I was left wondering if this incident was a joke or if the Captain had been serious. After all, some of our NCOs and Officers had tried the resolve and patience of our guard detail several times before since we had been assigned to our distant posts in these boondocks. One incident will bear that out.

We had been cautioned to never allow a piece of equipment to be removed from our post and especially an item of personal equipment. Losing our rifle was a Courts Martial offense and we were cautioned that several Marines that had been there before us had lost their weapons or had them taken from them while on guard duty. We were warned to be very alert and on our toes.

Early one morning our relief detail relieved me from my post at 0400. As our truck made the rounds to pick up the relief we picked up one Marine from his isolated post who had no rifle! He told the Sergeant of the Guard that someone had knocked him in the head and had taken his rifle and five round magazine of ammunition. He received a "dressing down" right there in the truck by the Sergeant who threatened jail time for the frightened Marine.

After arriving back at our barracks I got a chance to ask the Marine what happened. He explained that as he was walking post near a swampy area he felt a blow to his helmet which addled him. He fell to the deck and watched as two dark figures picked up his rifle and pulled his ammo magazine from his pouch. He said it was a few minutes before he was able to get up and call in the incident. He then looked at me and whispered, "I know who did it!" I looked at him quizzically and asked, " Who was it?" He answered, "It was two of our own NCOs!"

Sure enough it turned out to be two of our own Sergeant handlers who were making a point about the danger of losing our equipment. I know that I for one paid extra attention to my surroundings while on guard duty after that incident. I wasn't worried as much about losing my rifle as I was about getting knocked in the head!

Sunday evening at 1900 hours I was walking my regular post when I saw our guard detail pickup truck pulling into the parking area near me with the Sergeant that had posted me three hours earlier at the wheel. He had another Marine with him who I reconized as one of our regular guard Marines. The Sergeant approached me and said that he was relieving me from duty. Although his tone didn't seem harsh I couldn't understand why in the world I was being relieved from guard. I didn't question him as to the reason because Marines of lesser rank weren't in the habit of questioning a superior, being he an officer or an enlisted man.

As I seated myself in the front seat beside the Sergeant he looked at me and said," Private Nickels I have orders to bring you back to our Company area to see the Company Commander." I was really worried then as I realized that it would normally mean you had royally screwed up somewhere or you had an emergency at home if you were called before the Company Commander.

It took us about twenty minutes to reach the company area even though the Sergeant drove faster than normal. As soon as we came to a stop he told me to follow him to the Captain's office.

The Sergeant knocked on the Captain's hatch and I stood behind him as a voice from inside called out, "Enter!" We both entered the room and assumed the position of attention as the Sergeant reported to the Captain. After reporting, the Sergeant was dismissed by the Captain and I stood before him at attention until he looked at me and said, "At ease, Private!"

I noticed he had a worried look on his face and he hesitated as if he was trying to find the right words to say. After a minute of silence he said, "Private I'm afraid I have some bad news from home for you." As he spoke, my legs felt weak and I braced myself for the words that I knew must be bad and probably concerned one of my family members. He continued in a kind tone of voice , " I want you to know that I'm truly sorry for what I have to tell you but I've just received word that your Father has passed away. My driver is waiting in the outer office to take you to a phone booth where you can have some privacy and call home."

I was stunned by the news and could only stammer, "Thank you sir," as I snapped to attention, about faced and left his office. His driver, a Corporal,was waiting for me at the outer office hatch and motioned for me to follow him. We got into his jeep and drove to an area where there were about ten phone booths situated together.

There was a long lineof Marines and Sailors at each of the phones awaiting their turn to use the phone and when I saw them I dreaded having to wait in line to call home and find out what had happened. Dad had been in good health and had no major illnesses that I knew off when I had last seen him. I couldn't imagine what might have happened to him.

The Corporal told me to follow him and he surprised me by taking me to the head of one of the lines and announcing to the line of marines that I had an emergency at home and I had priority from the Company Commander to use the phone. The Marine talking on the phone heard what the corporal said and immediately hung up the phone and stepped out of the phone booth.

I dlaled the operator and placed a collect call to my Mom and Dad's home. Wanda was still staying with them as she had been doing since a week or so after I left home. I figured she would be the one that answered the phone when I called home that night.

Instead, the phone rang only once or twice before I heard the sound of the receiver being picked up. After a few moments hesitation I heard the familiar sound of my Dad's voice saying, "Hello!" I was shocked almost belyond belief to hear his voice. The thought raced through my mind at that instant that it must have been Wanda that something had happened to and that they had lied to me! I stammered out a questioning "Daddy?" He said, "Yeah, how're you doin'? Is everything alright?" I replied, "They told me you were dead. Is it really some one else that has died? Who is it? Tell me!" He said, "We're all fine here. What do you mean? Who told you I was dead?" I explained what I had been told and he said, "Well, they've made a bad mistake."

I was still not sure that they weren't keeping something from me so I interrupted him with, "Where's Wanda? Is she alright?" He said, "She's right here." I then heard him tell Wanda, "Here Wanda, talk to Eddie. They've told him one of us was dead!"

Wanda then got on the phone and I again went over what I had been told and how I had even been relieved from guard duty to call home. She assured me everyone was fine and even called Mom to the phone to speak to me. When Mom got on the phone she said that she and Wanda had tried to call me a couple of days earlier after not hearing from me for a few days and that they had finally reached someone from Camp Geiger that said they would try to get me word to call home. By the time that message reached me a couple of days later it had escalated into what was told to me by the Captain.

I explained to Mom that we still weren't allowed to use the phones unless it was an emergency so it would do no good to try to call me for a while. The reason I hadn't been writing often was because I didn't really have the time anymore and I wouldn't know my new address for a few days.

After I hung up I felt a whole lot better to know it was all a mistake but I felt embarrassed for all the trouble I had caused.* I told the Corporal to relay the information to the Captain when he returned to the Company area and to thank the Captain for me. Later, when I had time to analize what had occurred I felt that in reality I was the one that deserved the apology for all the stress I had suffered because of the mix up .

*My Dad lived for 26 more years after this incident, passing away in July, 1992. Ironically,in 1980,my mother had a heart attack and was admitted to the Appalachian Regional Hospital in Whitesburg, Kentucky. While the family was gathered there visiting her,she had another heart attack and was pronounced dead after C.P.R. and other procedures were unsuccessful in reviving her.

We were advised to make arrangments for her and were in the process of doing so, when a nurse rushed to the nurses station and shocked us by saying that Mom had miraculously started breathing again. She lived for 21 more years before passing away in October, 2001.I have always said that both my Dad and Mom died twice.

I never learned how a simple inquiring phone call from home had been transformed into something so much more serious by the time it reached me. Neither did I hear anything more about this incident from anyof my superiors afterwards.

On reflection, it was obvious that no single person was responsible for relaying such disinformation that night. Every Marine at every level had tried to be helpful to me even if their information had been so badly screwed up somewhere along the line.

It took me quiet some time to absorb all the mysterious incidents of that night. For days afterward I still wondered if A cruel joke had been played on me or if someone had just innocently screwed up somewhere along the line. I believe the latter is the correct interpretation of what happened.

Chapter Twenty

CAMP STONE BAY

Saturday, July 9, at 0800 my company was relieved from Base Guard Duty and transferred to Camp Stone Bay which was located several miles from Camp Geiger. Stone Bay was, like Camp Geiger, part of the Camp Lejeune complex and was located to the West of the New River Inlet which ran through the middle of the Greater Camp Lejeune Base area.

We were assembled on the street in front of our barracks and loaded into several 18 wheelers or enclosed troop carrier "cattle cars" which were used to transport our company to our new home. We were all anxious to begin ITR and consequently were in a jovial mood as we traveled through the dusty dirt backwoods roads. We thought we were in the boondocks while on guard duty but we discovered that our new camp was even more isolated from civilization than our former home.

The road we traveled was devoid of any buildings at all as far as I could tell, and the tall southern pine trees along the route seemed to be at least 100 feet high as I peered out the small windows of the troop trailer. Except for the flatness of the land the whole area reminded me of home where pine trees were never far from sight no matter where I had lived or worked in Kentucky. The red colored soil which was prevalent in this area helped to remind me that even though this area was similar, it still wasn't home.

Our new barracks was located among a large stand of pines but with very few other buildings in the area. The two lane road which led through the area was paved but there were several well traveled dirt roads which branched off into the dark woods and seemed to lead nowhere in particular. I assumed that these roads likely led to the different training areas we would be using as we advanced in our training.

Camp Stone Bay   Mon. Nite July 11

Hello Folks,

Well, I'm finally in ITR. Today was our first official training day. We'll be training seven days a week according to our Staff Sgt. Troop Handler. By the way, he seems to be a real decent man. He's a black Marine that has never, so far, raised his voice except when addressing us as a company! What a change!

We were issued our field equipment today so that we can have our first 10 mile forced march tomorrow. I hope the time on guard duty hasn't softened me. Ha! The worst thing about our new home is the heat and humidity. It seems almost twice hotter than what I experienced in South Carolina and that's saying something.

I'm slated to go straight to Lejeune after ITR so I probably won't be home for awhile.

We have fans and water in the barracks which makes it bearable, at least during the night. We have been sweating so much that our company are all losing weight. I literally stay wringing wet all the time. Maybe I can gain some weight after I return home. I won't be able to here though.

Tell Granny and little Kathy to write me when they can. I won't be able to call home from here because we can't use the phone except in emergencies.

I write Wanda every day but I don't know if the mail is regular from here. I haven't received any mail the last couple of days. It will take awhile to catch up with me. Camp Stone Bay is a nice place and the food is the best ever. We have to march or run through the woods for about 3 miles to reach the mess hall which is located beside the rifle range. The dust as we march makes it hard to breathe. Most of the time we have to run both ways! I better close for now. Love to all, Eddie

Our first day of training was taken up mostly with getting our equipment squared away but plenty of time was found for us to run and PT throughout the day. Our mess hall was located through the piney woods about three miles from our barracks area. Breakfast and evening chow was paid for in pain and suffering as we almost always double timed there and sometimes double timed back after chow. That little tidbit caused most of us to eat a light breakfast and supper. Lunch was usually served in the field. That is if you consider handing out c-rations to us by our Troop Handler as being served. I don't ever remember eating lunch in a mess hall while in ITR, although it might have happened on occasion.

Our normal routine would start each morning with 0400 hours being our normal reveille time. Then the three mile run to chow would be followed by the three mile return trip to the barracks. We would then "saddle up" for our day's training routine which would last until at least 2200 hours. We were literally training at least eighteen hours a day, including weekends. We were allowed to do our laundry at a building located nearby. This was one of the few buildings in the area and it had half a dozen washers and dryers we could use providing we had money. Otherwise our laundry was done in the head which wasn't attached to the barracks but was located nearby. The Marines would jump out of their racks at 0400 and tie a towel around their waists as they headed to the showers and the head. It wasn't too much of an inconvenience for us during my stay there but I imagine the winter months made traveling to the head pretty difficult to endure.

After our return from chow we would don our full packs, our rifles, bayonet, full canteen, web belt, first aid kit, two magazines, helmet, helmet liner, and sometimes be issued one or more C-rations for meals while in the field. We saw our living area few times during daylight hours the whole time we were in ITR.

Our sleep was limited to five hours at best due to the long days in training. It wasn't unusual at all to leave in the cattle cars at 0630 in the morning and return to the barracks area the next day at 0200 and catching a couple hours sleep before 0400 reveille rolled around.

While the stress level was lower here, sleeping while in class or while instruction on a weapons system in the Marine Corps inventory was sometimes hazardous to our health. We were in a Nuclear, Biological, and Chemical class one hot afternoon when several members of our company dozed off while the lecturer was droning on about the antidote for a certain poisonous gas. When he glanced up at the bleachers where we were seated he saw that even though it appeared everyone was looking straight at him, several Marines had their eyes closed and were snoring.

The lecturer, who happened to be an officer, screamed out at the offenders and called out each one to the platform beside him. When the 8 or 10 men were assembled and standing at attention beside him he handed each one a syringe and ordered each of them to demonstrate to the rest of us how to give ourselves the injection required for the antidote in the event of a chlorine chemical attack.

This antidote required the injection to be given in the thigh area of the body. It was almost comical to watch as the sleepy guilty ones pulled the plastic covers from the syringe and slapped the needles against their thighs to inject the serum. Every one of them grimaced like the pain was almost unbearable. It wasn't funny to them and it wasn't very funny to me as I knew it could have easily been me standing up there. Only the fear of possible consequences had kept me from dropping off to sleep also.

We had no time to make friends during training at ITR but my old Casual Platoon friend, Private Donald Millner, was in my Company and in my squad bay. We barely had time to say hello through the day as we were going in a run continuously. If we weren't running a field exercise we were in a classroom trying to study the nomenclature of the M1 rifle, the 3.5 rocket launcher, or some other weapon.

The M1 rifle was a wonderful weapon that was very much like the M14, except for the different size ammunition and the amount of rounds the weapon was capable of being loaded with. The M14 had a twenty round magazine loaded from the bottom of the receiver while the M1 was loaded from the top with an eight round clip. When inserting a new clip into the M1 Garand one has to bang the clip against the helmet to make sure the rounds of 30.06 ammunition are seated in the eight round strip clip evenly. If they're not, it's impossible to insert them for firing.

After a while it becomes second nature to remove an eight round clip of ammo from your bandoleer or web belt, then rap it sharply against the side of your helmet before inserting it into the top of the receiver. The M1 also gives off a distinctive *pinnng* sound when the empty clip is ejected. This created some problems when World War Two or Korean War participants used these weapons in combat. A nearby enemy sometimes knew by the sound of the clip being ejected when their rifle was empty and could take advantage of that fact while engaged in a battle.

We were deep inside the boondocks surrounding Camp Lejeune at one point in our training and were engaged in firing the 3.5 rocket launcher for familiarization with that weapon. It was still in use at that date in 1966 but was slowly being phased out in favor of a new weapon, a hand carried disposable rocket launcher which was capable of being carried and fired by one man. The 3.5" bazooka" required a two man crew, one to load and arm the weapon and one man to hold and fire it.

Although the bazooka had four times the range and its projectile weighed five times more, the new M72-66MM Light-anti-tank (LAW) was a welcome addition to the troops engaged in combat in the jungles of Vietnam. These weapons could take out a tank if need be but were used mostly as "bunker busters" in Vietnam.

We were engaged in firing our 3.5 bazooka at an old scrapped pair of tanks which were located about 200 yards away from our firing point. Each time the weapon was fired the back blast seemed to be (and was)as dangerous as the blast that would occur when the warhead exploded on the target.

When my time came to fire the bazooka my loader loaded the projectile and set the electrical arming device as I peered through the sight and aligned it with the target. When the loader tapped me on the helmet as a signal to fire when ready I took careful aim at a tank and pulled the trigger. I knew the back blast was awesome and I expected the weapon would have a gigantic kick but instead there was no noticeable recoil when I pulled the trigger of the bazooka. (The back blast was designed to disperse all the energy of the firing).

As the heavy round (5.5 kg) left the weapon the reduction of the weight from the back of the launcher caused me to slowly lean backwards while the weapon was still on my shoulder. I went so far back that the back of the launcher was seated against the deck while the muzzle was pointed vertically towards the sky! I thought the loader and the two nearby Instructors were going to die laughing. I'm glad I could give them a good laugh that day. I never got to see whether or not my rocket hit the target. I was too embarrassed to ask anyone. I hope I at least came close to the target.

One of the weapons we became familiar with during our infantry training was the M60 7.62MM General Purpose Automatic Machine Gun. This weapon could fire up to 550 rounds per minute and had a maximum range of over 4,000 yards. Its weight with attached tripod was approximately 16 lb and was worth its weight in gold, in my opinion.

When firing the M60 and watching the destructive power as the tracer and ball ammo smashes the target one wonders how anyone or anything can survive a hit from this weapon. Just the noise of the rounds being fired down range can be frightening.

Without doubt, the most fearsome and frightening weapon of all that we trained to use in ITR was the M2A1 Flamethrower. When I first fired it I was as afraid of firing it as any enemy might have been if the flaming mixture of gasoline, diesel fuel, and nitrogen had been aimed towards him. The word "awesome" is as good as any other to describe the fearful destructive power of this weapon.

The flamethrower consists of two small tanks with a short hose attached to them. It has two pistol grip triggers on the hose. The front trigger was the one that lit the match like device in front of the nozzle. The rear trigger released the napalm and the propellant from the two tanks spewed the burning mixture for a maximum distance of 30 to 40 yards.

This weapon served the Marine Corps well in the island fighting of World War Two, but was in the process of being phased our when I trained in the use of it during ITR. A weapon such as this might even be against the rules of war in today's world. I believe the shock of just seeing an enemy preparing to fire such a weapon in your direction would be terrifying in itself.

One hot July afternoon we began a ten mile forced march with our whole company of about 250 men who were already exhausted from our morning "fire and maneuver" exercises. I had filled my canteen up before I left the company area early that morning but had not had an opportunity to refill it when we started the forced hike with full field equipment, including helmet, rifle, bayonet, and eighty rounds of blank ammunition each which we were going to use in an exercise about halfway to our destination.

I had about one fourth of a canteen of water remaining at the start of the hike but I had drained it down to the last drop before we had covered the first mile. By the time we reached the area where we were going to assault a small knoll that had several log bunkers constructed on the sides and top of the knoll, I was almost desperate for something to drink. So was everyone else as their unanswered requests for water attested to. We were told that a "Water Buffalo"[1] would be along shortly and that the exercise must go on as the enemy "must be pushed from that hill!"

We split up into platoons and four man fire teams and using the fire and maneuver method of assault we captured the hill and its bunker complex. I managed to fire about 40 or 50 rounds of blank cartridges in my M1rifle without it jamming once. That was pretty unusual when using blanks.

Our assault and the discussion of our grading observers took a couple of hours to complete. As we prepared to resume our hike, my eyes (and others) searched almost desperately for a sign of a truck pulling a Water Buffalo coming up the dusty road

---

[1] A towed steel tank containing about 500 gallons of drinking water.

we had hiked in on. We searched in vain, for there was no truck in sight and the "show must go on."

We still had five miles to go and I began to search the roadsides and fields for a pool of water or a sinkhole with some water in it that I might quickly grab. We had received some rain over the last couple of days and I figured there must be some water somewhere in these hot, dusty piney woods that I could get a sip of.

A few Marines had a little water left in their canteens and shared their carefully conserved liquid as far as they could. They passed the canteens around as we continued marching along that extremely hot and dusty road. I was beginning to worry about heat exhaustion or heat stroke if we didn't get some water soon.

I was fortunate in being near the front of the formation where the dust wasn't nearly so bad as it was further back in the company's ranks. Being in front gave me another advantage. I was squinting my eyes and looking hundreds of feet ahead for a pool of water when I saw a small depression with a little water in it about five yards off the road and about 75 feet ahead of our column.

We were in route step which allowed me to quickly sprint to the side of the road, pull my canteen cup from the bottom of my canteen, and scoop up some muddy liquid into the cup. I turned it up to my lips and drank it down right there beside the road, mud and all. It was hot but it was wet and that's all that mattered at that moment. I now knew how those who are lost in the desert felt when they at last reached a pool of brackish water in the movies that saved their lives.

We lost several members of our company by the wayside that day due to the heat and lack of water. That one drink of muddy water sustained me until we reached our destination where a Water Buffalo was waiting to allow us to fill our canteens. I never did see a Water Buffalo coming up the road as promised.

A few days later we went on another forced march in the same area but we didn't suffer so much for water the second time around. Our forced march actually turned out to be a forced run, as when we had gone about a mile or so our troop handlers informed us that the last 25 men to cover the 10 mile hike would be put on mess duty for the next two weeks.

When we were told that, every Marine in the column took off on a dead run, leaving a cloud of dust behind them. I was no exception, as I sure didn't want to be cooped up in a hot scullery washing food trays at Camp Geiger if I could prevent it. One way I could prevent it was by making sure I did my 10 miles faster that at least 25 other Marines. That thought made me speed up even more, if possible.

As at Parris Island, my ability to persevere during our running exercises stood me in good stead this day. I started my eight or nine mile run by pacing myself and before too long I began to pass up many of those that had started the run at a faster pace. After a few miles the weight of all the equipment plus the M1 rifle began to feel as if they weighed as much as a 100 lb sack of horse feed. The heat and humidity caused all of us to have to pace ourselves eventually during the run to the finish line.

I was among the first 40 or 50 men to make it in to the finish line and I knew that I had escaped the dreaded mess hall. So did everyone else, including the last ones in. It turned out that the whole thing was a cruel joke being played on us. There was no mess duty at stake at all. Those Marines who were scattered all around the pines and bent over puking their guts out after the hot run didn't see one bit of humor in the exercise though.

We had all heard horror stories about the dreaded gas chamber from other Marines who had been through the CS gas chamber. It turns out they were mostly true to form, at least according to the experience my platoon had with the gas chamber.

We were taken to the igloo style aluminum building where our group would go through the dreaded gas chamber in the familiar cattle cars one hot July morning. We were led into the building in groups of about 30 Marines at a time.

My group watched other Marines exiting from the building before we had our turn. Some came out the door running, some crawling, and some were carried out. All of them had one thing in common. Their mouths were open and were frothy with a foamy stream of spittle and snot streaming down their chins and onto their utilities. They were rubbing their eyes vigorously as they coughed so hard that I thought they were in danger of rupturing themselves. Some ran for hundreds of feet while all this was occurring while others fell to the deck in apparent misery, rolling every which way along the grassy field.

Then, it was our turn. We put on our gas masks and entered the building where three NCOs in gas masks were gathered in the middle of the small room. The gas pellets had already been placed in a burner of some sort as we gathered in a circle around the room. We were ordered to adjust our masks as tight as possible by a Sergeant who seemed to be the one in charge of the exercise. His voice was muffled as he spoke inside his mask and we could barely make out his words.

We stood there as the gas became a fog which engulfed the room. The condensation of the eye pieces made seeing through the lens of the gas masks very difficult. After a few moments we were ordered to remove our masks and to sing the Marine Corps Hymn!

We had barely gotten through the first few words of the Hymn when the gas began to burn our throats and eyes so badly that we had to stop trying to sing while we coughed and tried to breathe. We were ordered by a muffled voice to "Put on your masks and clear them!"

We quickly put them on and cleared them but the remnants of the gas fumes inside the masks kept the coughing going as the fluids from our mouths and eyes poured into our masks and down our chins. I felt as if I wasn't getting enough air and what I was getting wasn't breathable.

One of our number was unable to get his mask successfully cleared and was in misery as he bent over retching and trying to gulp the chemically saturated air. After another 30 seconds he began kicking the sides of the aluminum building and hollering , "I can't clear my mask, let me out of here!" The three NCOs ignored the poor Marine as he continued to shout for the hatch to be opened. We had been warned that our masks might not be able to be cleared if they weren't the right size or if the straps weren't tightened properly.

What happened to that Marine that day was almost criminal, in my humble opinion. Tear gas is not a chemical that can be tolerated for even 30 seconds without collapsing in pain and misery. The Marine that was being ignored as he shouted that his mask wasn't working was left to suffer for nearly five minutes on his knees as he struggled to breathe. He had grown tired of kicking the side of the building after the ability to breathe had almost left him. After what seemed like several minutes someone finally opened the hatch and we streamed outside, retching our guts out like those before us. We had been ordered to remove our masks again for at least two minutes before the hatch was opened and our eyes were reddened and our mouths were dribbling foam as we had seen happen with the earlier group of Marines. I literally felt as though I might have ruptured myself coughing and retching.

The dreaded gas chamber was not something I was looking forward to after going through that first experience. Now I had a story to add to all the other "sea stories" I had heard about how difficult the gas chamber was to go through.

We headed to the grenade range one morning and spent the first part of the day in the classroom learning about safety procedures we would observe while our company was on the range. We were told some horror stories about how some men would freeze up after pulling the pins on a grenade and just drop the grenade in the concrete pit and take off running. The instructors emphasized the point that that the unexpected incident could be expected to happen when it came to throwing the grenade in an actual exercise.

They weren't kidding. As we took turns in the concrete pits we could see evidence of shrapnel hits all over the concrete pits. It was obvious that many grenades had exploded inside the pits instead of among the old tires situated about 25 yards downrange. I wondered in my mind how many had been seriously or killed while training to use this weapon.

It didn't take long for some of our group of Marines to screw up. Some managed to throw their grenades only 5 yards or less outside the pits because of the improper method of throwing them. It turned out that they were holding them too long before releasing them from their hands. Instead of throwing them like they would throw a baseball they were trying to "windmill" them. That resulted in a lot of close calls with shrapnel that day.

Our training at Infantry Training Regiment turned out to be even tougher physically than our time at Parris Island. We spent a portion of every day running for miles while carrying our total issue of equipment and our rifles. The physical exertion required for this phase of our training was unlike anything we had ever experienced. One thing was certain though. At the end of our four weeks, (including weekends), we all felt like we were ready for anything the Marine Corps could throw at us .

Chapter Twenty One

SUPPLY SCHOOL

On a very hot Friday afternoon in August, all 250 men of our Infantry Training Regiment class were standing in Company formation on the street in front of Camp Stone Bay barracks as a Marine Major congratulated us on our completion of ITR. He went on to say that we had completed two important phases of our training, (Boot Camp and ITR), and that our final stage, which would be our MOS schooling, was all that remained to be conquered and that he expected us to stay as motivated for the classroom as we were during the first two phases of training. He also said he hoped he might see each of us somewhere along the path of our Marine Corps journey. He called us to attention, gave us a salute, and dismissed us to prepare our gear to ship out to the Camp Lejeune main base later that afternoon.

A line of Marine Corps- green painted buses were waiting for us in the street as we packed our seabags and headed outside to board them. I boarded my assigned bus and sat in a window seat as I glanced around the Camp Stone Bay area that had been our home for more than four weeks. I had actually enjoyed our training in ITR; although I was sore all over and had many, cuts, scratches, and bruises that I had accumulated while spending so much time in the field. We had spent more nights in a tent and sleeping bag by far than we had spent in our assigned barracks. Unlike the one night I had spent at camp as a member of the Boy Scouts when I had nearly frozen to death, I had enjoyed my ITR time in the field on maneuvers. The much warmer weather might have made the difference for me.

We passed the entrance to the New River Air Facility and headed East on Route 17 and for the first time I saw the small city of Jacksonville, North Carolina, which was located only 7 or 8 miles from the Main Gate of Camp Lejeune. This town would become familiar to me in the next couple of years and would become almost like home to Wanda and I when she joined me there later.

We passed Jacksonville and proceeded along Lejeune Boulevard and soon came to the Main Gate near Midway Park. We were saluted as we passed the white- gloved sentries and then continued along the four lane high way until the buildings of the sprawling Marine Corps Base came into sight.

The bus made several turns before the brick barrack buildings came into sight for the first time. As the bus driver made our first stop he called the name of a Marine from his list and said to him, "This is where you get off. Report to that building, (pointing to a large brick building), which is the Field Music School for Basic Musicians. So you're gonna be a Bugler huh?" The Marine answered in the affirmative and I watched in awe as he hoisted his seabag and exited the bus for what I considered to be a pie job. I had trained with this particular Marine and he had never spoken about what his assigned job would be during our conversations we had had in the field. I think that if I was slated to be a bugler I would have announced it to the world and not kept it to myself. A bugler could be expected to be assigned to Sea Duty at some point in their career and the chance for promotion was thought by some to be much greater in that career field.

We soon came to a group of buildings where our bus pulled over and my name was among the several Marines the Driver called out. He pointed to a building which he identified as H&S Company, Base Material Battalion, which was the company all the General Warehousemen, (MOS 3051), were assigned to. This was to be my new home for several weeks.

When I checked in to H&S Company I found that it would be almost two weeks before Warehousing School started and that I would be sent on leave for 10 days to await the start of school. I wasn't given any choice in the matter but I sure wasn't going to argue with the First Sergeant that gave me the order to go on leave. Besides, I was ready for a rest after staying in the boondocks for almost five weeks during ITR.

A Corporal of H&S Company led me to a storage room where I could stow my miscellaneous gear and seabag while on leave, as I would be assigned to another barracks for Supply School when I returned from leave. I then waited in the recreation room with a couple of other Supply Marines who would also be going on leave.

I picked up my leave papers and we were given a ride to the bus station which was in the central part of the base, and rode the bus to Jacksonville where I bought a ticket to Jenkins, Kentucky. I had just been paid before leaving Camp Stone Bay or I couldn't have possibly bought even a bar of candy that day. The ticket wasn't very much at all but I can't recall the cost of it.

When I boarded the Trailway bus I found myself having to stand up in the aisle because of the overcrowded conditions. The only luggage I had with me was a small duffel bag with a few items in it which I had handed to the bus driver to stow in the luggage compartment as I climbed aboard the bus. Standing up wasn't a big problem for me and I figured a seat would open up when someone got off the bus somewhere along the line.

I had changed into my summer uniform when at the H&S Barracks as I didn't have any civilian clothes to wear home. We were never allowed any leisure time or liberty during our training and I had no use for casual clothing. Also, if one traveled in uniform a discount of your ticket price was given you. I needed every discount I could get as my pay check for two weeks was only $43. 95 and I would have to share as much as I could with Wanda. Her allotment checks had been sporadic at best and I believe she had only received one check at that point.

I rode standing up all the way from Jacksonville to Raleigh, N. Carolina before we pulled into the bus station for a short break. I found myself a seat inside the station while we were waiting and while I was sitting there I was approached by a middle aged man who shook hands with me and thanked me for my service. He struck up a conversation about the Marine Corps and he remarked to me that "I wore the Eagle, Globe, and Anchor during World War Two and fought the Japs!" I was impressed with him after those words and thanked him for his service also.

About that time the bus to Winston Salem, North Carolina and Abingdon and Bristol, Virginia was announced over the loudspeaker. As I rose to leave, my new friend wished me good luck, then asked if I had a quarter to spare! I realized then that he was a pan handler that all bus stations of that era seemed to be plagued with. I made a mental note to be more careful in selecting friends on the rest of my journey. (I gave him a dollar.)

As I think back, the man never said that he was in the Marines during World War Two. He had merely said that he had worn the Eagle, Globe and Anchor. He probably bought them at a military surplus store. I was learning fast that this world could be a very strange place. He more than likely did his fighting in his mind or in a bar. I guess I'm a skeptic at heart.

I found that the bus was once again filled to over capacity and I had to stand in the aisle once more as we headed towards home. No one offered this poor, tired, lonely Jarhead a seat. Ha! I'm joking of course, as it was every man or woman for them-selves.

We changed buses in Winston Salem after we spent a few hours of waiting to make connections. Again I watched as people almost fought for a place near the bus doors as we lined up for boarding. Again I was unable to get a seat. I know that wartime almost always means more crowed buses, planes, and trains but this was becoming ridiculous, I thought.

It was after 0100 hours in the morning before our bus pulled into Bristol, Virginia after stopping in Abingdon for a couple of hours. We had another long wait for the bus to Jenkins, which wasn't scheduled to leave Bristol until 0630 that morning. My legs felt like jelly after having to stand all the way from the East Coast of North Carolina to Bristol, Virginia while standing and holding onto a seat corner to keep from falling as we rounded curvy roads. I had the consolation of knowing that at least I wouldn't have to stand on the last leg of the trip. The route between Bristol and Jenkins was unlikely to be as crowded as our route through the state of North Carolina. Or so I thought.

As we boarded the bus to Jenkins early that morning, I was mortified to find that I would once again be forced to stand in the center aisle. I was tempted to use some of my newly found muscles acquired in training to grab the arm of some unsuspecting elderly lady and sling her into the aisle while I sat down in her seat! I never thought too much of grabbing a man and taking his seat. Not that there's anything wrong with being an elderly lady! I never felt like trying the old adage that one Marine could easily whip four men. So I once again grabbed the edge of a seat and leaned this way and that as the choking diesel smoke rolled throughout the older model bus that had been assigned to the Jenkins route. I had my doubts of it being able to make it up and down the many steep inclines without the engine conking out. I for one would have been glad for the reprieve from diesel fumes for a little while.

We arrived in Jenkins at around 0900 that morning and I phoned Dad's house and gave him a surprise when I told him I was in Jenkins. I hadn't called before leaving Camp Lejeune and no one was expecting me to get a leave before finishing Supply School.

As I waited for Dad to come from Whitesburg to pick me up I stood outside the bus station and reflected on what I had been through since leaving home in March.

I thought back to how I had stayed up for four days and three nights before being allowed to sleep when first entering Parris Island, how I had come close to being shot at the rifle range while carrying the Lieutenant's raincoat to him, how I had suffered a broken foot, how while at the Base Dispensary I had been involuntarily photographed (naked) to have my stress fracture injury included in a medical book, how my lungs had been seared with CS gas, how I had been beaten, kicked, strangled, forced to eat a plate of spaghetti, and suffered other humiliations during boot camp, how I, along with many others, had been strapped by a wild NCO with a whip or cat-o- nines tail while standing formation in Camp Geiger, had nearly been frozen to death and came near to heat stroke during boot camp, how I had nearly died from thirst during a forced hike, how I had to help disarm a knife wielding recruit that was trying to harm himself and others, and how the many bus rides I had taken since entering the draft had nearly smothered me to death with diesel fumes. Other than that my life had been pretty much uneventful since entering the Marine Corps.

At that very moment an elderly man came around the corner and walked towards me. He then said to me that he had seen me standing there from the nearby gas station and just wanted to thank me for my service. He then asked me, "How do you like the Marines?" Without missing a beat I replied, "Oh, I love it!" May the Lord forgive me for that little white lie. I'm sure that was the answer he expected to hear and I didn't want to disappoint him by telling him the whole truth!

My leave time was spent mostly in just relaxing around the house at Dad's with only a visit to the Alene Theater with Wanda for entertainment. Evenings were spent in reminiscing about the past and discussing the possibility of Wanda moving to North Carolina to live with me while I completed my schooling. Although my future after school was still unknown we felt that it wouldn't be too difficult for her to move back to Kentucky if I was sent overseas after school ended. Besides, she didn't want to be a burden on my family anymore.

We decided to load up a few belongings in Dad's car which he insisted I take back to Camp Lejeune.(He didn't have to twist my arm for me to agree to the offer.) Wanda and I left a couple of days early for Lejeune to try to find an apartment to rent. I was pleasantly surprised to find that she had saved most of her allotment money she had received which would allow us to rent a place without having to borrow money.

When my leave was almost over we loaded the car with some of Wanda's belongings and drove the 525 miles to Jacksonville and rented a motel room for Wanda near the entrance to New River Air facility. She stayed there for nearly two weeks until we found a garage apartment in the middle of the city. Several Marines and their families also rented apartments in the nearby large apartment building which was part of the complex. We became good friends with a couple from Georgia, Robert and Sue Sumpter.

Robert was also attending the Supply and Administration School at Camp Lejeune. After over four weeks of schooling

301

which consisted of morning classes and afternoons working in the large warehouses of Camp Lejeune, our company of 150 Marines graduated from Supply School with a MOS of 3051.

This was a MOS of General Warehouseman, which was considered one of the better jobs in supply. A supply platoon could always be identified during an inspection by the amount of new 782 accouterments lying on their racks for their regular "Junk on the Bunk" inspections. These inspections occurred every three months and much preparation went into the placement of each piece of equipment and each item of uniform on the rack in exactly the order the book called for.

The rifle and its accouterments were part of this quarterly inspection and were expected to be in tip top condition. To fail a Junk on the Bunk was to lose liberty privileges and would likely result in extra duty. A re-inspection was required within 24 hours of a failed inspection. Very few Marines had to have more than one re-inspection. He or she usually got serious after the first failure.

One particular inspection we stood Wanda prepared my clothing and equipment for inspection and boxed it all up so all I had to do was take it on base and lay it out on my rack for inspection. I passed the inspection with an "Excellent" grade while unfortunately my friend Robert failed his and had to retake the inspection the next day. He was peeved at me because he knew Wanda had done my work for me. He told all our fellow supply Marines that "Eddie's wife passed the inspection, not him!" He was sometime getting over this little incident. The fact that I only smiled when he mentioned it didn't help to make him feel any better.

We graduated from Supply School early on a Friday morning and that afternoon we were loaded into the familiar "cattle cars" of an eighteen wheeler which would drop each of us off at our assigned permanent duty station on Camp Lejeune. I had found out just before graduation that I would be stationed on Camp Lejeune and would be assigned to the Second Marine Division. I didn't yet know exactly what regiment, battalion, or company would be my new home as I boarded the familiar cattle cars that afternoon.

The truck driver made several stops throughout the base, letting one Marine off at one company of the 6th Marine Regiment, and three Marines off at another company of the same regiment. A Sergeant was in the truck trailer with us with a manifest and would call out the names at each stop.

We left the 6th Marine Regiment area and stopped in the street in front of the first barracks we came to in the 2nd Marine Regiment Area. The sergeant checked his list and called out, "Privates Eddie G. Nickels and Robert Sumpter, this is where you two get off. The barracks in front of you is the location of Headquarters Company, Second Marine Regiment, Second Marine Division. This will be your new home. Report promptly to the First Sergeant in the company office immediately upon entering the building. Good luck to both of you!"

I now knew where I would be working as a Warehouseman. To the surprise of us both, Robert Sumpter and I, who were already good friends that lived in adjoining apartments off base in Jacksonville, were being assigned to the same outfit and would be doing the same job. He and I could hardly believe our luck.

We were both in similar circumstances even before our assignment to the same company occurred. He and I had been drafted, we had both took a chance by bringing our wives to Camp Lejeune not knowing where we would end up after school, we both attended the same school and classroom in Supply School, we had an apartment in the same complex in Jacksonville, and had been given the same MOS. The greatest difference between us was our ages. He was twenty five years old and I was nineteen years old. He was older than the average new Marine, even at the tender age of twenty five.

His wife Sue and my wife became the best of friends during our time at Lejeune. Wanda still didn't drive so she and Sue often did their errands together with Sue driving. Wanda and she did their grocery shopping at the PX on base and would often visit us at work when they were on base.

The date I reported to Headquarters Company, Second Marines, Second Marine Division, was September 16, 1966. I had left home in Kentucky on March 23 and had been in training for 26 weeks before I was assigned to my first permanent duty station. It had been a long journey in which I had learned a lot about human nature and about the country which lay outside the Appalachian Mountain Area I had been born and raised in.

The six months training I had undergone in order to became a United States Marine had changed me as a person. I was no longer that shy teenage boy that didn't have an inkling what life was all about. I now felt that I could persevere no matter how difficult life in the future might be.

The almost impossible to accomplish training situations I had endured over the last six months had shown me that one could overcome almost all obstacles of life by using discipline, willpower, and perseverance to achieve any worthwhile goal in life. All the other attributes learned while going through Marine Corps training had one common denominator needed to reach those worthy goals. That common denominator was confidence. Without confidence, no goal is achievable. I give proper credit to our Parris Island Drill Instructors for installing that confidence into my mindset. The ways of our Drill Instructors were tough, demanding, and yes, sometimes even brutal, but in the end, their training methods brought out the best in all Recruits.

I will always be proud to my dying day to carry forever the title of that which I worked so hard to achieve, the title of United States Marine.

Semper Fidelis.

Chapter Twenty Two

EPILOGUE

My time in Headquarters Company, 2nd Marines, was an interesting period of time because of all the activity going on in Vietnam. Some members of our Headquarters Supply unit were Temporary Additional Duty (TAD) infantrymen recently returned from a combat tour in Vietnam and they had many interesting stories about their tours there. One thing I found in common among them was the feeling that we could have won the war at any time we wanted to if the American political leaders had allowed our military to fight an offensive war. They were all frustrated by the fact that they weren't allowed to be aggressive in their tactics while serving there. Putting your troops into enclaves and being on defense is no way to win a war for Marines that have been trained to be aggressive and to attack, not hide in a camp surrounded by barbed wire obstructions. We all felt we were not being used to win a war, but merely to make a political point. What a way to fight a war.

I spent only a month as a Warehouseman before my Commanding Officer, Lt. Fecke, moved me into the office and made me his Fiscal Requisition Clerk. I was in charge of equipment inventory and of ordering parts and equipment for our Company. I enjoyed the job and even the responsibility it entailed. It made me feel that I was contributing to the war effort when I ordered field jackets, C- rations, or a truck part that enabled a battalion of Marines to ship out for a world trouble spot. Most of our battalions of the 2nd Marines were rotated on cruises, either to the Caribbean Sea or the Mediterranean Sea.

I received orders for Vietnam in October, 1966 but because our supply unit was so short handed the Lt. requested that my orders be postponed or cancelled. He spoke to me in private and explained that my orders had been postponed for the present. In February, 1967, I again received orders for Vietnam and I made arrangements to get Wanda settled back in Kentucky before I had to leave. I was scheduled to leave for additional training on the following Monday. On the Friday before, a fellow Marine, Lance Corporal Thoms, who was from Utah, asked me if he could try to volunteer to go to Vietnam in my place. He had over two years remaining to serve and he said, "I'll have to go eventually, and it may as well be now rather than later." I was already dreading to go on account of having to take Wanda home to Kentucky that weekend and his offer was a Godsend to me.

We went straight to Lt. Fecke's desk and asked if a switch would be possible? After a pause he replied, " I'm not sure but I'll make some quick calls." At 1530 hours he called Thoms and me to his desk and told us that he had received permission to substitute L/Cpl. Thoms for myself. Talk about a reprieve at the eleventh hour! Our normal quitting time was 1600 hours. Thirty minutes later I would have left for home in Midway Park and then left for Kentucky with Wanda. I've always felt that God's hand was guiding our lives for sure after this incident. I had never minded going to Vietnam but we weren't financially able for Wanda to have to go back to Kentucky at that time.

Wanda and I enjoyed the eighteen months we lived In Onslow County. North Carolina, as we were fortunate to live only a few miles from the Atlantic Ocean. We spent many weekends at the nearby beaches which normally had few people using them during those days. Some weekends we would go to the Surf City Beach and find that we were nearly the only ones on the beach! Try that today.

Our son Steven was born during our military tour, on November 14, 1966. He was born at the Naval Hospital located on Camp Lejeune. The Navy doctors and that hospital were among the best that could be found anywhere in the world.

Before his birth we had moved from Jacksonville to Midway Park, North Carolina. This put us several miles closer to the hospital. When it came time for Steven to be born, Wanda's good friend Sue rushed her to the Naval Hospital which was located on a rise of land near the New River Inlet. Lt. Fecke let me off from work that morning so I could be there for his birth.

 Our military time was a time of hardship because of my low pay. At my highest rank, Lance Corporal, my pay check for two weeks was $62.40. Rent, food, taxes, gas, and entertainment had to come out of this. Before becoming pregnant Wanda worked at Rose's Department Store on Lejeune Boulevard which helped us out tremendously for a while. She couldn't drive so Sue or I had to take her to work and back. That arrangement eventually became too much for everyone concerned and she had to stop working.

I received my release from active duty on Friday, March 22, 1968 and was transferred to the 47th Rifle Co., FMF,USMCR, (4) Louisville, Ky., for the remainder of my obligated service.

Four years later, March 23, 1972, I received my Honorable Discharge from the United States Marine Corps. I had finally successfully completed my six year military obligation that had begun with my trip from Whitesburg, Ky. to Ashland, Ky. to be inducted into the Marine Corps six years before on March 23, 1966. The unforgettable memories made during those years will last a lifetime for Wanda and me.

On November 24, 1969 our daughter Alisha was born and on July, 30, 1972, our son Jeffrey arrived in the world. We now have eight wonderful grandchildren and four wonderful great-great grandchildren. I couldn't ask for anything more to verify our blessings and success in life.

APPENDIX

Military Time

a. Has no colon to separate hours and minutes.
b. Hours are numbered 1 through 24 instead of using a.m. and p.m.
c. Has no designation "o'clock."
d. A zero precedes the hours 1 through 9.

0100	Zero one hundred	1:00 a.m.
0200	Zero two hundred	2:00 a.m.
0300	Zero three hundred	3:00 a.m.
0400	Zero four hundred	4:00 a.m.
0500	Zero five hundred	5:00 a.m.
0600	Zero six hundred	6:00 a.m.
0700	Zero seven hundred	7:00 a.m.
0800	Zero eight hundred	8:00 a.m.
0900	Zero nine hundred	9:00 a.m.
1000	Ten hundred	10:00 a.m.
1100	Eleven hundred	11:00 a.m.
1200	Twelve hundred	12:00 noon
1300	Thirteen hundred	1:00 p.m.
1400	Fourteen hundred	2:00 p.m.
1500	Fifteen hundred	3:00 p.m.
1600	Sixteen hundred	4:00 p.m.
1700	Seventeen hundred	5:00 p.m.
1800	Eighteen hundred	6:00 p.m.
1900	Nineteen hundred	7:00 p.m.
2000	Twenty hundred	8:00 p.m.
2100	Twenty-one hundred	9:00 p.m.
2200	Twenty-two hundred	10:00 p.m.
2300	Twenty-three hundred	11:00 p.m.
2400	Twenty-four hundred	12:00 midnight
0005	Zero Zero Zero Five	12:05 a.m.

Glossary of Terms

Ashore	Off station. Where you go on leave or liberty
Aye Aye, Sir	Official acknowledgement of an order
Barracks	Building where Marines live
Below	Downstairs
Bivouac	An area where you pitch tents in the field to stay overnight
Blouse	Coat
Boondocks	Woods or wilds; training area
Brightwork	Brass or shiny metal; i.e., water faucets, doorknobs, etc.
Bulkhead	Wall
Bunk or rack	Bed
Chit	A small piece of paper, a receipt or authorization
CMC	Commandant of the Marine Corps
CO	Commander, Commanding Officer
Colors	A national flag
Cover	Hat
Cruise or tour	Period of enlistment
Deck	Floor
Drill	March
Esprit de Corps	Spirit of camaraderie
Field	Boondocks where you train
Field day	Clean up an area
Galley	Kitchen
Gangway	Move out of the way or make room
Gear Locker	Storage room or locker for cleaning purposes
Gung Ho	Working together; in the spirit
Hatch	Door
Head	Bathroom

Ladder . . . . Stairs
Leave . . . . Authorized vacation
Liberty . . . . Authorized free time, but not leave
MOS. . . . . Military occupational specialty
NCO . . . . Non commissioned officer
NCOIC . . . . Noncommissioned officer in charge
Overhead . . . . Ceiling
Passageway . . . . Corridor or hallway
Porthole . . . . Window
PFT. . . . . Physical fitness test
PX. . . . . Post Exchange; comparable to a civilian
                          department store
Quarters . . . . A place to live; i.e., house, barracks, etc.
Reveille. . . . . Time to get up
Secure. . . . . Stop work, put away, close or lock
Scuttlebutt . . . . Water fountain, rumors
Snapping in . . . . Practicing getting into firing position
Squadbay . . . . Large room in barracks where Marines live
Square away . . . . Straighten up, make neat
Survey. . . . . Turn in unserviceable items
Swab . . . . Mop
Taps . . . . Time to sleep
Topside . . . . Upstairs
W.M. . . . . Woman Marine

Marine Enlisted Insignia and Pay Grades

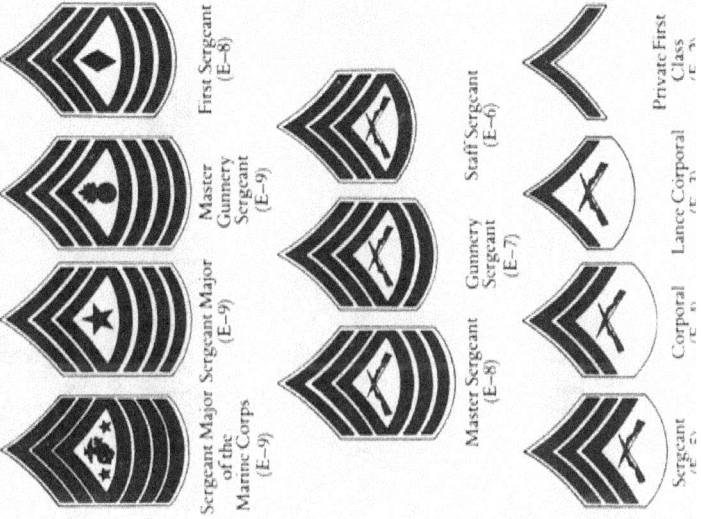

First Sergeant (E-8)

Master Gunnery Sergeant (E-9)

Sergeant Major (E-9)

Sergeant Major of the Marine Corps (E-9)

Staff Sergeant (E-6)

Gunnery Sergeant (E-7)

Master Sergeant (E-8)

Private First Class (E-2)

Lance Corporal (E-3)

Corporal (E-4)

Sergeant (E-5)

Marine Officer Insignia and Pay Grades

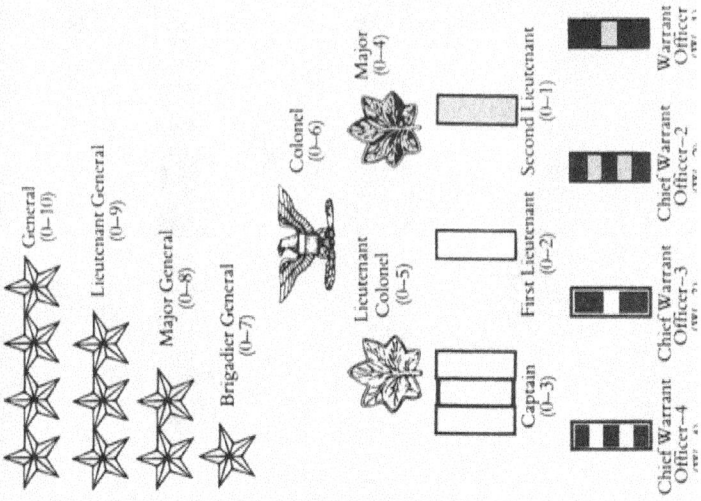

General (O-10)

Lieutenant General (O-9)

Major General (O-8)

Brigadier General (O-7)

Colonel (O-6)

Lieutenant Colonel (O-5)

Major (O-4)

Captain (O-3)

First Lieutenant (O-2)

Second Lieutenant (O-1)

Warrant Officer (W-1)

Chief Warrant Officer-2 (W-2)

Chief Warrant Officer-3 (W-3)

Chief Warrant Officer-4 (W-4)

SELECTIVE SERVICE SYSTEM

Approval Not Required.

ORDER TO REPORT FOR
ARMED FORCES PHYSICAL EXAMINATION

Local Board No. 58
Selective Service
Post Office Building
Whitesburg, Kentucky 41858

(LOCAL BOARD STAMP)

To   Eddie Gregory Nickels
      P. O. Box No. 215
      Whitesburg, Kentucky 41858

24 January 1966
(Date of mailing)

SELECTIVE SERVICE NO.			
15	58	47	215

You are hereby directed to present yourself for Armed Forces Physical Examination to the Local Board named above by reporting at:

Local Board No. 58, Rm. No. 1, U. S. Post Office Bldg, Whitesburg, Ky.
(Place of reporting)

on   9 February 1966   at   5 AM EST
           (Date)   (Hour)

Lelia H. Banks
(Member or clerk of Local Board)

IMPORTANT NOTICE
(Read Each Paragraph Carefully)

TO ALL REGISTRANTS:

When you report pursuant to this order you will be forwarded to an Armed Forces Examining Station where it will be determined whether you are qualified for military service under current standards. Upon completion of your examination, you will be returned to the place of reporting designated above. It is possible that you may be retained at the Examining Station for more than 1 day for the purpose of further testing or for medical consultation. You will be furnished transportation, and meals and lodging when necessary, from the place of reporting designated above to the Examining Station and return. Following your examination your local board will mail you a statement issued by the commanding officer of the station showing whether you are qualified for military service under current standards.

If you are employed, you should inform your employer of this order and that the examination is merely to determine whether you are qualified for military service. To protect your right to return to your job, you must report for work as soon as possible after the completion of your examination. You may jeopardize your reemployment rights if you do not report for work at the beginning of your next regularly scheduled working period after you have returned to your place of employment.

IF YOU HAVE HAD PREVIOUS MILITARY SERVICE, OR ARE NOW A MEMBER OF THE NATIONAL GUARD OR A RESERVE COMPONENT OF THE ARMED FORCES, BRING EVIDENCE WITH YOU. IF YOU WEAR GLASSES, BRING THEM. IF MARRIED, BRING PROOF OF YOUR MARRIAGE. IF YOU HAVE ANY PHYSICAL OR MENTAL CONDITION WHICH, IN YOUR OPINION, MAY DISQUALIFY YOU FOR SERVICE IN THE ARMED FORCES, BRING A PHYSICIAN'S CERTIFICATE DESCRIBING THAT CONDITION, IF NOT ALREADY FURNISHED TO YOUR LOCAL BOARD.

If you are so far from your own Local Board that reporting in compliance with this Order will be a hardship and you desire to report to the Local Board in the area in which you are now located, take this Order and go immediately to that Local Board and make written request for transfer for examination.

TO CLASS I-A AND I-A-O REGISTRANTS:

If you fail to report for examination as directed, you may be declared delinquent and ordered to report for induction into the Armed Forces. You will also be subject to fine and imprisonment under the provisions of the Universal Military Training and Service Act, as amended.

TO CLASS I-O REGISTRANTS:

This examination is given for the purpose of determining whether you are qualified for military service. If you are found qualified, you will be available, in lieu of induction, to be ordered to perform civilian work contributing to the maintenance of the national health, safety or interest. If you fail to report for or to submit to this examination, you will be subject to be ordered to perform civilian work in the same manner as if you had taken the examination and had been found qualified for military service.

SSS Form 223 (Revised 4-28-65) (Previous printings may be used until exhausted.)   ☆U.S. GOVERNMENT PRINTING OFFICE: 1965—O-774-926

Code of Conduct

Article I

I am an American fighting man. I serve in the forces which guard my country and our way of life. I am prepared to give my life in their defense.

Article II

I will never surrender of my own free will. If in command, I will never surrender my men while they still have the means to resist.

Article III

If I am captured, I will continue to resist by all means available. I will make every effort to escape and aid others to escape. I will accept neither parole nor special favors from the enemy.

Article IV

If I become a prisoner of war, I will keep faith with my fellow prisoners. I will give no information or take part in any action which might be harmful to my comrades. If I am senior, I will take command. If not, I will obey the lawful orders of those appointed over me and will back them up in every way.

Article V

When questioned, should I become a prisoner of war, I am required to give only name, rank, service number and date of birth. I will evade answering further questions to the utmost of my ability. I will make no oral or written statements disloyal to my country and its allies or harmful to their cause.

Article VI

I will never forget that I am an American fighting m responsible for my actions, and dedicated to the princi which made my country free. I will trust in my God and in United States of America.

Marines' Hymn

From the Halls of Montezuma,
To the shores of Tripoli,
We fight our country's battles
In the air, on land and sea.
First to fight for right and freedom,
and to keep our honor clean;
We are proud to claim the title of
UNITED STATES MARINE.

Our flag's unfurled to every breeze
From dawn to setting sun.
We have fought in every clime and place
Where we could take a gun.
In the snow of far off northern lands
and in sunny tropic scenes,
You will find us always on the job,
The UNITED STATES MARINES.

Here's health to you and to our Corps
Which we are proud to serve.
In many a strife we've fought for life
And never lost our nerve.
If the Army and the Navy
Ever look on Heaven's scenes,
They will find the streets are guarded by
UNITED STATES MARINES.

315

AUTHOR'S NOTE

While all the recollections, events, incidents, and participants mentioned in this book are composed of factual events and real people, I have taken the liberty of changing the names of some of those participants in order to protect their privacy.

Some information used came from letters exchanged with the family during my time at Parris Island and some other dates and events were from notes and documents I have saved from that period.

Any errors or mistakes in this book are mine and mine alone and can be attributed to the number of years that have passed or to my poor recollections.

I have kept the dirty language used in boot camp to a bare minimum to only show the flavor of what was said and have included none of the cursing except for some abbreviations where I needed to express some tensions as they happened. I could never feel comfortable even writing down the language as used on most occasions at that time in boot camp. I understand that today's training at Parris Island is conducted without the bad language and without the level of intensity of earlier years. The young men and women training there today still uphold the finest traditions of one of the World's best fighting forces.

www.ingramcontent.com/pod-product-compliance
Lightning Source LLC
Chambersburg PA
CBHW051940090426
42741CB00008B/1209